Made for Each Other

Made for Each Other

A SYMBIOSIS OF BIRDS AND PINES

Ronald M. Lanner

New York Oxford
Oxford University Press
1996

Oxford University Press

Oxford New York
Athens Auckland Bangkok Bogotá Bombay
Buenos Aires Calcutta Cape Town Dar es Salaam
Delhi Florence Hong Kong Istanbuul Karachi
Kuala Lumpur Madras Madrid Melbourne
Mexico City Nairobi Paris Singapore
Taipei Tokyo Toronto

and associated companies in
Berlin Ibadan

Published by Oxford University Press, Inc.,
198 Madison Avenue, New York, New York 10016

Library of Congress Cataloging-in-Publication Data
Lanner, Ronald M.
Made for each other : a symbiosis of birds and pines
/ Ronald M. Lanner.
p. cm. Includes bibliographical references.
ISBN 0-19-508902-2 (cloth). — ISBN 0-19-508903-0 (pbk.)
1. Jays—Ecology. 2. Pine—Ecology.
3. Animal-plant relationships. I. Title.
QL696.P2367L35 1996
598.8'64—dc20 95-45844

Frontispiece: The North American corvids that harvest, eat, and cache pine nuts. Clockwise from top—Clark's Nutcracker, Pinyon Jay, Scrub Jay, Steller's Jay. Sketch by Claudet Kennedy.

1 3 5 7 9 8 6 4 2

Printed in the United States of America
on acid-free paper

For William Morehouse Harlow
(1900–1986)
dendrologist; master teacher; cinematic pioneer;
warm friend; lover of the great American forest and
the lore of its early inhabitants.

ACKNOWLEDGMENTS

While researching and writing this book, I was helped by many students, professional colleagues, family members, and friends. The enlightened policies of my college and department administrators—Thad Box, Joe Chapman, Dick Fisher, Charles Grier, and Terry Sharik—gave me the time I needed to reconstruct what I had learned over a seventeen-year period. Bartell Jensen provided funds for an important journey to Finland, where Veikko Koski put me in touch with Teijo Nikkanen, who in turn put me in touch with migrant nutcrackers. My field studies in Squaw Basin, Wyoming, were ably assisted by Karen Snethen, Steve Ruddell, David Lanner, and Harry Hutchins; and by personnel of the Bridger-Teton and Shoshone national forests. Hutchins continued to make important information available to me long after his student days, most recently after he and his wife Sue had made the most thorough field studies to date of the Korean stone pine in its natural habitat. During the period of my research, I was helped immeasurably by H. Charles Romesburg on matters statistical and philosophical; by Barrie Gilbert, on bears; Russ Balda and the late David Balph, on bird behavior; Gail Fondahl and Marina Tolmacheva, on Siberian lore; Carling Malouf and Nancy Turner, on Salish Indian uses of whitebark pine nuts; John Gurnell, on squirrels; Konstantin Krutovskii, on Siberian stone pine; Shinichiro Saito, on the Japanese stone pine; and Walter Schönenberger, Rudolf Häsler, and Reudi Zuber, on the stone pines of Switzerland. Fondahl graciously translated key portions of several Russian articles. Claude Crocq generously shared with me his knowledge of European nutcrackers and directed me to their haunts in the French Alps. Long conversations with Kate Kendall about her bear research, and with Diana Tomback about her nutcracker work and numerous other aspects of the corvid-pine mutualism, increased my understanding and sharpened my perceptions. Bud Cheff graciously shared his memories of a boyhood rich in whitebark pine experiences. Other contributions came from Hermann Mattes, Friedrich-Karl Holtmeier, Ward McCaughey, Connie Millar, Tad Weaver, Steve Arno, Dave Mattson, Yan Linhart, Paul Zedler, Ray Hoff, Fred Wagner, Gene Schupp, Gary Belovsky, Ben Munger, Harold D. Picton, Doug Ramsey, and Kathleen Braden. My wife Harriette has my gratitude for enriching so many field trips, for encouraging me in my research, and for providing valuable insight. Cogent comments on the manuscript were made at one stage or another by Tom Lyon, Jillyn Smith, Barrie Gilbert, Kate Kendall, Harry Hutchins, Steve Vander Wall, and Connie Millar. For precise and efficient word processing I thank Melanie Brunson, Francene Rasmussen, Carolyn Brooks, and Lana Barr. Finally, I will always be grateful to Steve Vander Wall for sharing with me for two decades his encyclopedic knowledge of corvid biology; and to my great friend, the late Bill Critchfield, who taught me much of what I know about pines, and gave me the incentive to learn the rest.

CONTENTS

Made *for* Each Other

CHAPTER 1

Introduction

OME TREES AND BIRDS ARE MADE FOR EACH other. Take for example the whitebark pine, a timberline tree that graces the moraines and ridgetops of the northern Rockies and Sierra Nevada-Cascades system (Plate I.1). This lovely five-needled pine, long-lived and rugged though it is, cannot reproduce without the help of Clark's Nutcracker. And the nutcracker, though it captures insects in the summer and steals an occasional bit of carrion, cannot raise its young in these alpine habitats without feeding them the nutritious seeds of the whitebark pine. Between them, these dwellers of the high mountains provide for each others' posterity, which leads biologists to label their relationship symbiotic, or mutualistic. But there is more to it than that, because in playing out their roles these partners change the landscape. The environment they create provides life's necessities to many other plants and animals. Working in concert, Clark's Nutcracker and the whitebark pine build ecosystems.

The seeds and cones of the whitebark pine seem at first glance to be useless. The cones do not open far enough to let the seeds fall out. But even if they did, the big, heavy seeds would not fly gracefully to new habitats the way those of most other conifers do, because they lack the membrane wing that grants mobility. Left undisturbed, whitebark cones are not shed from the tree until the following summer, at which point the seeds are rancid and unable to sprout. Truly, the whitebark pine seems boxed in, painted into a reproductive corner from which there is no escape. What in the world is it doing here?

The nutcracker, too, seems at first to be misplaced. Visit a pine grove at the winter solstice, and you will hear the throaty squawks of resident Clark's Nutcrackers. Yet no cones remain in the treetops, and neither fruits nor crawling creatures can be found. Temperatures are now below zero, drastically increasing the birds' metabolic needs when food is at its scarcest. What in the world are these nutcrackers doing here? The best response to such vexing questions is to have a closer look.

Nutcrackers start to feed on whitebark pine seeds in mid-August. The tightly closed purple cones are pulpy, and the nutcrackers shred them with their chisel-like bills to remove the seeds. By September the cones are dry and brown, and the scales have loosened. They have become brittle and are easily broken from the axis. The broken-off scales fall to the ground, but their seeds remain firmly held in the core of the cone. The nutcracker removes these. Some are shelled and eaten immediately upon being harvested, but most are dropped into a sublingual pouch, an opening in the floor of the mouth beneath the bird's tongue. Only nutcrackers, members of the genus *Nucifraga*, have such pouches. A full pouch can hold more than eighty whitebark pine seeds.

The nutcracker now flies to its cache site. It brings up the pouched seeds into its bill, one by one, and thrusts them into the soil, about an inch beneath the surface. Then it hops off and makes other caches nearby until its pouch is emptied. A cache may contain only one seed, or as many as fifteen. The seeds are in no way damaged during the harvesting or caching process. Seeds may be cached within a hundred yards of the trees they came from, or up to several miles away. Many seeds are first cached nearby and later transferred to more distant locations. Seeds are often buried in open areas and in recently burned-over forests. Many caches are made where wind keeps the snow swept clear and winter access is assured, but nutcrackers will retrieve seeds from beneath the snow if they must.

Nutcrackers start to recover their caches in the fall, while they are still busily harvesting and caching; they live off the cached seeds until a new seed crop appears the following summer. Experiments have shown convincingly that nutcrackers find their caches by relying on memory—they *remember* where each group of seeds has been buried. It appears that what they specifically remember are the angles between their caches and certain nearby landmarks, like boulders, trees, stumps, and logs: in other words, they triangulate. This is a remarkable finding for several reasons. First, it implies long-term memory, because some caches made in September may not be unearthed until the following July, ten months later. Also, snow modifies the landscape, obscuring many of the landmarks the nutcrackers must use as cues. Finally, the task requires remembering so very much, because research indicates a single bird may conceal as many as ninety-eight thousand seeds in over thirty thousand caches. Perhaps only half of these will be recovered and eaten by the nutcracker that cached them, or fed to its young in the spring.

When summer comes to the high country the moisture of melting snow and the heat of the warming sun quicken those seeds still buried in the soil. With luck, many of them will germinate successfully and will end the summer with long taproots embedded in the drying earth. In a few years a small grove of whitebark pines will mark the caching area used by nutcrackers in the recent past. Many of the pines in these groves are in clumps of two to eight stems or more, reflecting the number of seeds that germinated from the same cache. The large seed is advantageous during the establishment process, because its ample reserves of stored food give the new seedling a good start. The burial of the seed reduces losses to predaceous rodents, and protects it from the drying sun. As the grove develops and enlarges through germination of seeds that nutcrackers continue to cache there, spruces start to appear.

Engelmann spruce prefers some shade for its proper development. If it grows in the open sun, an Engelmann spruce seedling will become "solarized"—that is, it will suffer chlorophyll breakdown and needle death. But the half-shade of the pine grove is a habitat friendly to young spruces, and soon the grove becomes a mixed forest. The shade cast by these trees and the water they transpire into the air change the local microclimate. It becomes cooler in summer, warmer in winter, and less windy at all times. These conditions encourage further tree growth, and eventually the patches of open-grown woodland coalesce, forming a continuous forest.

Larger forest stands offer cover for large animals like bear, elk, and moose, and provide territories for Red Squirrels. These squirrels are especially important to the economy of the developing forest. They depend largely on conifer seeds for a food source, and beginning in late summer they cut cones of all species from the trees. These are stored on the ground, piled up against fallen logs or stumps, or buried in the forest litter. Such "midden" areas are used year after year, and deep within them one can often find decaying cones stored long ago but still bearing healthy seeds. The cones are removed as they are needed during the long winter, the scales gnawed off, and the seeds eaten.

The accumulation of high-energy food in a midden full of whitebark pine cones becomes a temptation more than most bears can bear. In the fall both Black and Grizzly Bears are preparing to hibernate and must increase their stores of body fat. The seeds of whitebark pine are large and very rich, containing 50 to 60 percent fat, and are an ideal food for this purpose. So the large beasts enthusiastically plow up the squirrels' middens, crush the cones underfoot, and gingerly extract the delectable seeds. A good whitebark cone crop is an important asset to the bears, and attracts them to these high-elevation forests.

As the forest spreads, its canopy shades out some of the sun-loving plants that dominated the open ground. Sagebrush, grasses, Indian paintbrush, fireweed, and bistorts disappear. Their place is taken by plants of the half-shade like grouse whortleberry, gooseberry, helianthella, and parrot's-beak pedicularis. Seeds of subalpine fir drift in, and soon firs join the pines and spruces in the developing forest.

Some trees die young and become thin cylinders of dead wood that decay rapidly when they fall. Others, like old whitebark pines, are killed in bark beetle attacks that leave them standing a century or more as bleached columns of resinous wood. Woodpeckers, nuthatches, and other cavity-nesters take up residence in these snags, and wood-rotting fungi and bacteria break them down chemically. Other fungi enter mycorrhizal associations with conifer roots. These fungi form the mushrooms that are their fruiting-bodies and which are eagerly devoured by Mule Deer and Red Squirrels in search of proteins. Dark forms glide in and out of the forest as Pine Martens and Steller's Jays take cover in its shadows. In the fall, when the conifer seed crops mature, finches, siskins, and crossbills forage in the spruce-tops. Grosbeaks, Steller's Jays, and even an occasional raven join the nutcrackers in the whitebark pine seed harvest, though only the nutcracker takes significant quantities. Gray Jays fly secretively into the pines and spruces, concealing bits of carrion among the branches.

The forest of whitebark pine and Engelmann spruce becomes self-perpetuating. Spruce and pine seedlings in large numbers stand ready to grow into the canopy when space is made available by the death of a mature tree. Barring severe disturbance, this forest appears capable of remaining stable for long periods of time. Stability, however, is seldom realized in nature, and before many decades have gone by, the forest is likely to be ignited in one of the thunderstorms so common and so violent in the high country. A catastrophic fire creates an open area attractive to nutcrackers in need of an open cache site, and the cycle begins again.

The whitebark pine-Clark's Nutcracker relationship fits the common definition of a symbiosis or mutualism: the coexistence of a plant and an animal in a mutually beneficial partnership. Pine-corvid symbioses range across much of the Northern Hemisphere and include numerous pines and corvid species. Despite the fact that most natural history enthusiasts, biologists, foresters, ornithologists, and botanists are unaware of this phenomenon, the dispersal of pine seeds far from the mother tree by nutcrackers and jays is easily observed.

And it happens on a grand scale. The symbiosis characterizes over twenty species of pine, including some of the most widespread, and is the majority condition among the world's so-called soft pines. These corvid-dispersed soft pines—referred to in this book as bird pines—are immensely important to the ecology of vast tracts of North American and Asian wildland. On this continent, for example, seventy-five thousand square miles of terrain in the southwestern United States is dominated by pinyon pines of several species. The nuts produced by these woodland-forming pines, the very seeds that attract Pinyon Jays and other corvids, were once a mainstay of Native Americans. Related species still serve as an important food source for the inhabitants of millions of hectares of rural Mexico. In the Rockies, the Cascades, the Sierra Nevada, and numerous ranges of the Great Basin, ecologically sensitive areas at high elevations are protected by whitebark or limber pines, both nutcracker-dependent species. Like the pinyon pines, whitebark and limber pines are distributed entirely within the geographic range of "pinivorous" jays and nutcrackers (Figure 1.1). The Western landscape owes much to these feathered cultivators of our rocky highlands.

The Alps and Carpathians form a relatively tiny European locus for the pine-corvid symbiosis. But further east, over the Ural summits, stretches a nutcracker-established forestscape of continental scope. The enormous resource provided by the Asian stone pines—close relatives of the whitebark pine—has seldom been recognized. According to Chinese authorities, a fourth of the timber now being cut in that sprawling land is Korean stone pine. The vastness of the Asian heartland that is clothed in Japanese and Siberian stone pine almost defies imagination. The former spans over thirty degrees of latitude; the latter stretches across six time zones. There are winter days when the sun sets on one forest of Siberian stone pine while it rises on another. It was only partly in jest that I wrote a few years ago that a hungry nutcracker could conceivably eat its way through four thousand miles of stone pines from the Urals to the Bering Sea.

These are interesting facts of natural history, and they satisfy our constant craving to know how nature works. But beyond that, the story of the pine-corvid symbiosis has scientific weight as well. Its magnitude compels the attention of plant ecologists and biogeographers. The impact of small birds on the landscape, given large numbers, determination, and plenty of time, is massive. In these corvid-established forests, a tree grows where a bird planted it: purposiveness, not randomness, is the organizing principle of structure. Consider also the match of arboreal and avian traits: cones that ac-

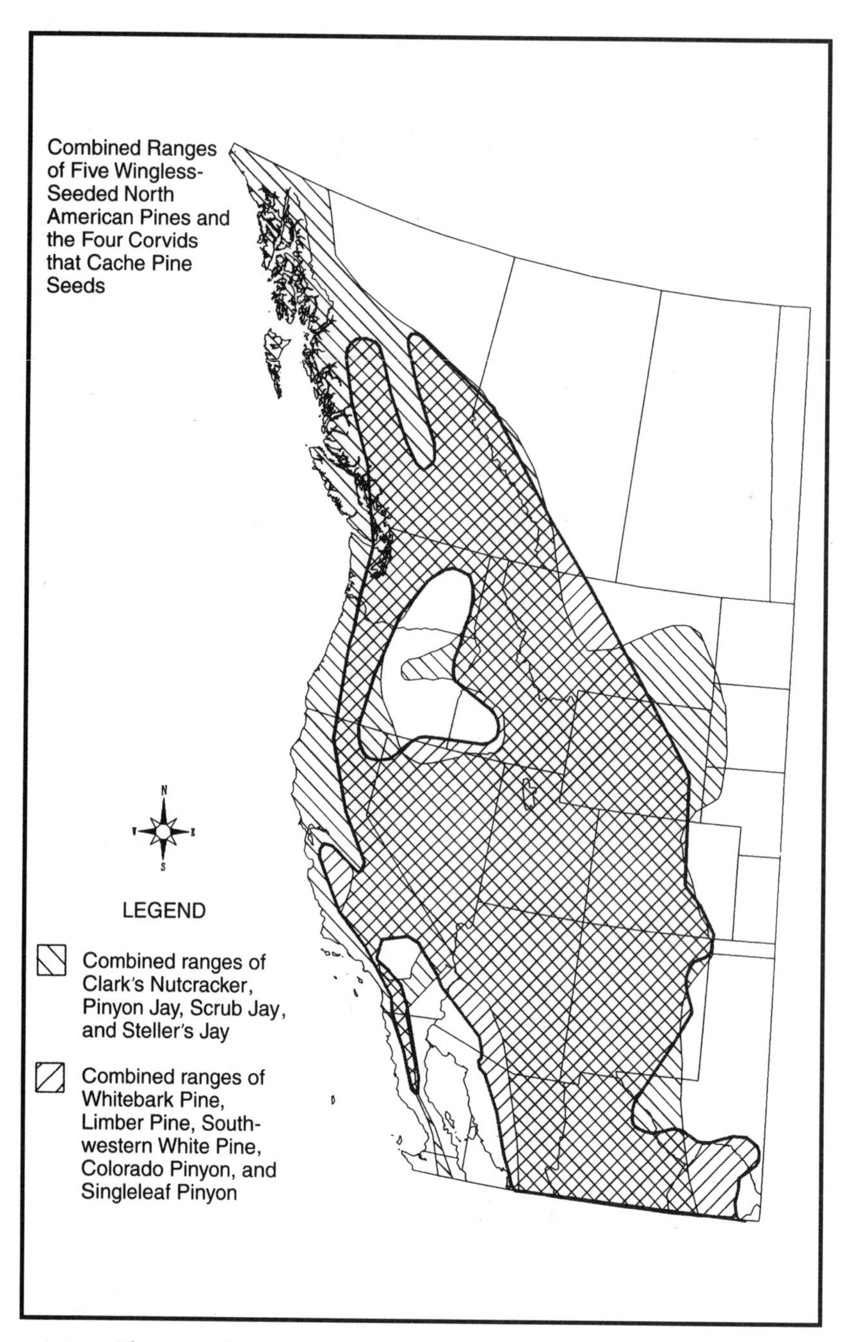

FIG. 1.1 There is a close correspondence between the ranges of the wingless-seeded soft pines of western North America and the corvids known to harvest, disperse, and cache their seeds.

commodate bills, seeds that fill dietary needs, branches that display cones to searching eyes, dormancy that promises future sustenance. Are such congruencies mere coincidence? Or are they evolved, as the actors in mutualism play against one another? In other words, have these pines and their corvid dispersers *coevolved*? If so, has it happened more than once? Does it continue? What does the future hold?

These questions cannot be fully answered by case studies like those presented in the pages that follow. But they can be illuminated and more sharply defined. And the knowledge gained may help in averting disaster for the whitebark pine, the most prominent of the trees discussed in this book, which is threatened by a disease without a cure in addition to other troubles. These, then, are the objectives of this book: to describe compelling natural history while raising scientific consciousness, and to raise an alert.

To keep the text from becoming overly cluttered, I have relegated personal crotchets, technical esoterica, and bibliographic details to the notes section following chapter 14.

CHAPTER 2

The Genus of Pines

LAPP REINDEER HERDSMAN APPROACHING Rovaniemi. A Oaxacan campesino piling firewood sticks on his burro. A turpentiner from the Landes of Gascony, an Evenki truck driver, an Alabama short-order cook. A vintner of Ravenna, a Harbin dentist, Scottish harpers playing old ballads in a minor key. A seemingly diverse lot with little in common to bind them? Perhaps. Yet all were probably raised, if not in the shadow, then at least within resin-scent or pollen-flight of the trees we call pines. The baobab may have been, as Peter Matthiessen would have it, the tree where man was born; but it was at the foot of the pine tree that man came of age.

More than a hundred pines—species of the genus *Pinus*—encircle the northern hemisphere in a great variety of habitats. One tropical pine breaches the Equator in Sumatra. Several hardy species extend north of the Arctic Circle. Pines grow from sea level to treeline, at desert edge and in rain forests. As beautiful as butterflies, as resourceful as cockroaches, pines have survived and flourished for 130 million years, since the early Cretaceous Period of the Mesozoic Era.

Pines are distinguished from all other cone-bearing trees by adult foliage made up of fascicles, or bundles, of evergreen needles, wrapped at least temporarily at the base in a sheath of bud-scale remnants, the "fascicle sheath" (Figure 2.1). The number of needles per fascicle, usually two to five, is quite consistent within each species. In this respect pines differ from the younger relatives with which they are grouped in Pinaceae, or the pine family. These include the hemlocks (*Tsuga*), Douglas-firs (*Pseudotsuga*), spruces (*Picea*),

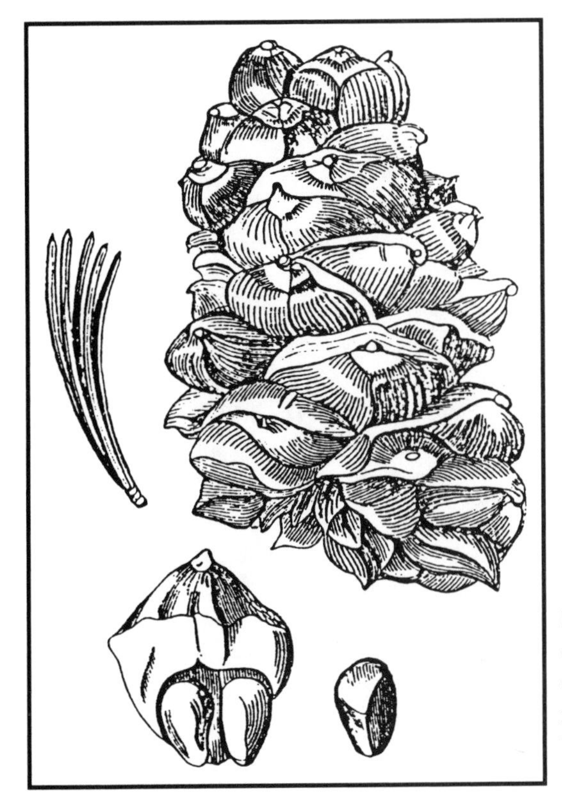

FIG. 2.1 Whitebark pine cone, cone-scale with seeds, seed, and needle fascicle, as illustrated by Thomas Nuttall for Newberry's report (1857).

firs (*Abies*), and little-known *Keteleeria* and *Cathaya* of China, all of which have needles attached singly to the branchlet, as well as larches (*Larix*), golden larch (*Pseudolarix*), and true cedars (*Cedrus*), that bear most of their foliage in many-needled tufts emanating from peglike woody spurs. Pine needles may be only three-fourths of an inch long (in jack pine) or as long as a foot and a half (longleaf pine).

They are pleasantly fragrant trees, due largely to the essential oils that evaporate through their needle stomates, especially during hot weather, and to the resin they exude from wounds and which bleeds from growing cones. Most pines are "early seral" in ecologist's jargon—that is, they are among the first trees to grow on open ground that is being naturally revegetated, and they tend to form pure stands in which most of the trees are about the same age. They are considered sun-loving and shade-intolerant because they usually do not reproduce well in the shade of an overstory. Therefore they are often transients on the land, being eventually replaced by the more shade-tolerant trees that grow beneath them. They can remain present, however, for a long while. Their thick bark, tolerance of heat and drought, and seeds that germinate best on bare soil, allow many pines to grow successfully in frequently burned areas. Pines are thus able to dominate landscapes where the average interval between fires is relatively short.

Pines are of immense economic importance over much of the world because of their wood, which is easily worked, strong, light, and long-fibered, and which is used for a myriad of lumber products and paper-pulp. The rapid growth of many species and their adaptability to growth in plantations also contributes to their significance. Many pines grow to large size. They are usually among the tallest trees wherever they are found; most species reach heights of over one hundred feet, and a few even attain two hundred feet. The largest of the genus is the sugar pine, which has been known to grow as tall as two hundred fifty feet, and has left at least one stump over thirteen feet across in the Sierra Nevada.

Pines are locally valuable for the production of turpentine and other naval stores from the resin of their trunks, and for pine nuts, their large, edible seeds, which will provide the dynamic for the story told in this book. Pines have long been of great cultural significance wherever they grow, as a source of legend and inspiration for painters, poets, and musicians. Much of the veneration of pines may stem from their relative longevity: a pine is the only organism known to have attained four thousand years of age. They are considered a symbol of long life in the Far East. Finally, pines are individually picturesque, reflecting in their forms the vagaries of environment and experience more freely than rigid firs and spruces.

About two-thirds of the pines are classified in the subgenus *Pinus,* once called the Diploxylon pines, and commonly referred to as hard or yellow pines. They differ consistently from the Haploxylon—soft, or white pines of subgenus *Strobus*—by having two veins running the length of the needle instead of one. They also differ by having harder and yellower wood (due to a high summerwood content in the annual ring); cones that are armed with bristles, prickles, spines or hooks; and rougher (scalier) young shoots; but these characters show some overlap. Fascicle sheaths of hard pines are almost always permanent, while those of soft pines usually disintegrate within a year. Hard pines often have stiff, shiny foliage, described by Donald Culross Peattie as "burnished like metal . . . a star of sunlight blazes fixedly in the heart of each strong terminal tuft of needles." Typical North American hard pines include ponderosa and lodgepole in the West, longleaf and loblolly in the Southeast, and pitch and red pines in the Northeast. Scotch and Austrian pines, among the most commonly planted ornamentals in the northern United States, are hard pines.

Hard pines will play a minor role in this book, because only four of the sixty-odd species of this subgenus have seeds that are known or suspected to be harvested and dispersed by corvids. The Italian stone pine, or umbrella pine, which lends a characteristic flavor to Mediterranean landscapes, has large seeds with ineffective wings, and is almost certainly dispersed by Azure-winged or Common Magpies. (Ironically, despite its occurrence near some of the western world's oldest seats of learning—the classical universities of Spain, Provence, and Italy—the natural history of this striking tree is largely unknown.) Three California species—Digger (gray), Coulter, and Torrey pines—have very large, thick-shelled seeds with stubby wings, equally unlikely to be effective as instruments of flight. Scrub Jays are known to take their seeds, and to cache those of the Torrey pine, but whether these cached seeds germinate and become seedlings is not known. Steller's Jays are also potential dispersers of these pines, and the possibility of dispersal and establishment by rodents cannot be ignored. Ponderosa and Jeffrey pine seeds are frequently taken by Clark's Nutcrackers, and by Steller's and Pinyon Jays, but apparently not on a scale large enough to have affected the biology of these trees.

And so, by default, the story of the relationship of corvids and pines must center around the subgenus *Strobus*—the soft pines. The close affinity of corvids to soft pines is due to the much higher frequency of wingless or ineffectively winged seeds in that subgenus: two-thirds of the thirty-odd soft pines have been denied the possibility of flight.

What exactly are seed wings, and what is their origin? The seed wing is a wedge of membranous tissue attached to the blunt outer end of the seed (Figure 2.2 a, b). It may be as short as six millimeters in jack pine, or as long as fifty millimeters in longleaf pine. It is often light brown or silvery tan in color, semi-transparent, and may be six times as long as the seed it is attached to. Wings extend toward the outside of the cone, where, upon parting of the scales, they can be lifted by the breeze. The wing causes the seed, when airborne, to spin, or autorotate. Autorotation slows a seed's descent, increasing the opportunity for winds to carry it a long distance. Thus wings are functional structures that facilitate the dispersal of potential offspring away from the mother tree's zone of influence. Wind dispersal is further aided by the placement of the cones high in the treetops, and by the small size and light weight of the seed in relation to the size of its wing. A seed that is generously winged is therefore well-adapted to wind dispersal. But a heavy pine seed with a rudimentary wing (Figure 2.2 c), or none at all (Figure 2.2 d, e), can only be expected to fall like a stone into the inhospitable pool of shade at the base of its mother tree. Such a seed can be effectively dispersed away from its

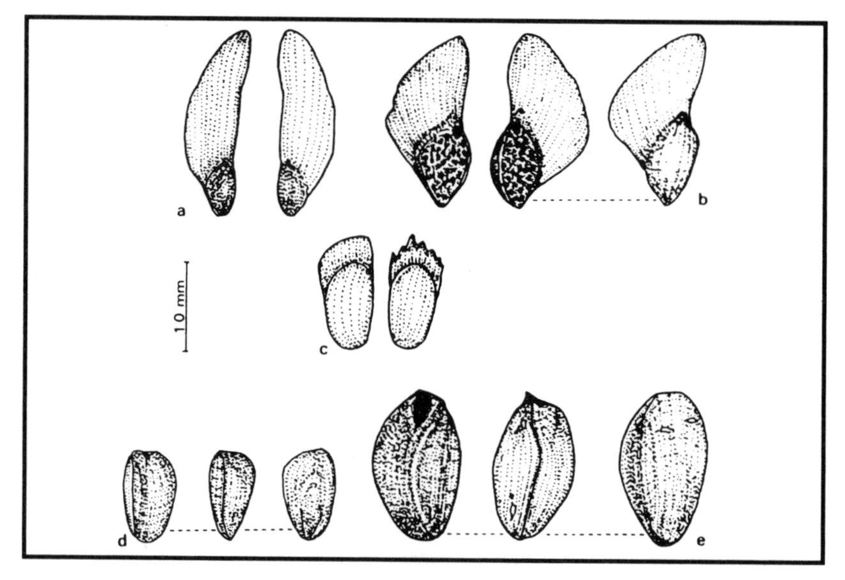

FIG. 2.2 Winged, wingless, and nearly-wingless seeds of soft pines: (a) eastern white pine; (b) Japanese white pine, var. *pentaphylla*; (c) Japanese white pine, var. *himekomastsu*; (d) Japanese stone pine; (e) Korean stone pine. Sketches by S. Saito, courtesy of Shiretoko Museum.

origin only by an animal agent. Therein lies the biological significance of the wingless seed, which makes its species hostage to the needs and desires of birds or mammals.

All of the five groups of soft pines—subsections according to Little and Critchfield's classification—are touched by the relationship of wingless-seeded pines and avian dispersers; and all contain species that will be discussed later in some detail. Here is a quick survey of these groups, the individual species of which are listed in Table 2.1.

THE *BALFOURIANAE*, OR FOXTAIL PINES

This small group—consisting of Great Basin bristlecone pine, Rocky Mountain bristlecone pine, and foxtail pine—is strictly North American, and confined to high western mountains. Its best-known species is the Great Basin bristlecone, whose notoriety stems from the good luck of some trees to have lived over four thousand years. All of these have small winged seeds. However, because of its proximity to wingless-seeded pines important to Clark's Nutcracker, the Great Basin bristlecone has itself beome prey to the high-

Table 2.1
The "soft pines," *Pinus* subgenus *Strobus*.[a]

Scientific name	Common name	Generalized distribution
Section Parrya		
Subsection *Balfourianae*	foxtail pines	
Pinus balfouriana	foxtail pine	California
P. aristata	Rocky Mountain bristlecone pine	Colorado, New Mexico, Arizona
P. longaeva	Great Basin bristlecone pine	California, Nevada, Utah
Subsection *Cembroides*	pinyon pines	
*† *P. cembroides*	Mexican pinyon	Mexico, SW U.S.
*† *P. edulis*	Colorado pinyon	Southwestern U.S.
*† *P. juarezensis*	Sierra Juárez pinyon	California, Baja California
*† *P. monophylla*	singleleaf pinyon	SW U.S., Mexico
*† *P. culminicola*	Potosí pinyon	Mexico
*† *P. maximartinezii*	Martínez pinyon	Mexico
*† *P. pinceana*	Pince pinyon	Mexico
*† *P. nelsonii*	Nelson pinyon	Mexico
*† *P. johannis*	Zacatecas pinyon	Mexico
*† *P. discolor*	border pinyon	SW U.S., Mexico
*† *P. remota*	Texas pinyon	Texas, Mexico
Subsection *Gerardianae*	Asiatic nut pines	
*† *P. gerardiana*	chilgoza pine	Indian subcontinent
*† *P. bungeana*	lacebark pine	China
Section Strobus		
Subsection *Strobi*	white pines	
P. strobus	eastern white pine	Eastern North America
P. monticola	western white pine	NW North America
P. lambertiana	sugar pine	California & vicinity
* *P. flexilis*	limber pine	Western North America
* *P. strobiformis*	southwestern white pine	SW U.S., Mexico
* *P. ayacahuite*	Mexican white pine	Mexico, Central America
P. peuce	Balkan white pine	Balkan Peninsula
* *P. armandii*	Armand Pine	China, Tibet, Burma
P. griffithii	Himalayan blue pine	Indian subcontinent
P. dalatensis	Vietnamese white pine	Vietnam
*† *P parviflora*	Japanese white pine	Japan
P. morrisonicola	Taiwan white pine	Taiwan
P. fenzeliana	Fenzel pine	China, Vietnam
P. wangii	Yunnan white pine	China
Subsection *Cembrae*	stone pines	
*† *P. albicaulis*	whitebark pine	Western North America
*† *P. cembra*	Swiss stone pine	Europe
*† *P. sibirica*	Siberian stone pine	Siberia, Mongolia
*† *P. pumila*	Japanese stone pine	Northern Far East
*† *P. koraiensis*	Korean stone pine	Northern Far East

Source: Adapted from Critchfield and Little (1969), but see notes to chapter 2.
[a]An asterisk (*) indicates species with wingless or almost wingless seed in at least one variety. A dagger (†) indicates a species with cones that retain their seed after maturation.

flying corvid. This unexpected relationship will be examined in some detail in chapter 12. The name "foxtail" is taken from the long lengths of heavily needled branch tips that characterize these closely related pines.

THE *CEMBROIDES* OR PINYON (PIÑON) PINES

These Mexican and southwestern United States species have been especially important as providers of human food. They all grow in semi-arid country and are typically broad-crowned and of low stature. The group is evolutionarily active, with frequent hybridization among some of its species, and is in some taxonomic disarray. Thus some botanical "splitters" would argue there are more species than I have listed in Table 2.1, and some "lumpers" would complain I have included too many. Of the five species found in the United States, the most widespread are Colorado and singleleaf pinyons. All the pinyon pines have totally wingless seeds which are held in the open cone by growths of cone-scale tissue (Figure 2.3). This characteristic is unknown in other pines, and is perhaps the defining morphological characteristic of this ecologically and geographically coherent group. *Cembroides* includes a single-needled species, the only such pine; a two-needled species; and several

FIG. 2.3 An open cone of singleleaf pinyon showing the rims of cone-scale tissue (light crescents) retaining the large wingless seeds. Utah State University Photography Service.

three-and five-needled species. *Cembroides* is named for one of its members, *Pinus cembroides,* the Mexican pinyon, which received *its* name because of a fancied resemblance to *Pinus cembra* of Europe.

THE *GERARDIANAE* OR ASIATIC NUT PINES

These two species—chilgoza pine in the Himalaya and Waziristan, and lace bark pine, found at scattered locations in China—superficially resemble the pinyon pines. The former is an important nut provider; the latter is frequently planted in the Far East to display its beautiful sycamore-like bark. Both species have reduced seed wings that cannot function in wind dispersal because they remain stuck to the cone scales (Figure 2.4). This was noted by Shaw as early as 1914, and I have verified it in a large number of chilgoza pine cones, and several of lacebark pine. While the wing of a chilgoza pine seed sticks to the lower surface of the scale above it, that of a lace-bark pine seed sticks to the upper surface of its own scale. Both are three needled. Chilgoza pine, *Pinus gerardiana*, is the namesake of this group.

FIG. 2.4 Cone of lacebark pine showing the narrow seed wings adhering to the cone-scale surface and retaining the seeds. Utah State University Photography Service.

THE *STROBI* OR WHITE PINES

This largest of the *Strobus* subsections is something of a catch-all for species that do not fit elsewhere. Most of them are magnificent forest trees distinguished by neatly layered tiers of horizontal branches clothed in fine-textured, blue-green needles. Their clusters of pendant, finger-shaped cones tend to have long-winged seeds that are easily carried by the wind. These "conventional" white pines have been important sources of high-quality lumber in North America and Eurasia. But a few of the *Strobi* pines are scraggly little trees with limby crowns bearing cones containing big, heavy, wingless or nearly wingless seeds. Of these, only the Japanese white pine, *Pinus parviflora*, retains some of the seeds in its cones (Figure 2.5). Like lacebark pine, its seed wings stick to their scales. All these pines are five-needled. As we will see, several of the *Strobi* pines yield clues that are highly suggestive of the evolution of the pine-corvid symbiosis. "*Strobi*" is derived from *Pinus strobus*, the eastern white pine.

THE *CEMBRAE* OR STONE PINES

The stone pines have in common five-needled fascicles, wingless seeds, and cones that remain closed indefinitely even after the seeds have matured (Fig-

FIG. 2.5 Japanese white pine cone showing tip of the seed-wing adhering to the cone-scale surface. The seeds are becoming detached from the wings, which will remain in the cone after nutcrackers harvest the seeds. Utah State University Photography Service.

ure 2.1). They are Swiss stone pine (*Pinus cembra*), Siberian stone pine (*Pinus sibirica*), Japanese stone pine (*Pinus pumila*), and Korean stone pine (*Pinus koraiensis*), in addition to our own whitebark pine (*Pinus albicaulis*). (The Italian stone pine, despite its name, is one of the hard pines and is unrelated to the "true" stone pines of subsection *Cembrae*.) Many other soft pines—indeed most of them—are five-needled; and seed winglessness is common enough as well: what really sets the stone pines apart is their peculiar non-opening cone. The nature of this cone, and its implications for both pine and corvid biology, will be explored in detail in later chapters.

Swiss stone pine is widely distributed high in the Alps of Switzerland, Austria, France, Germany, Slovenia, and Italy. Further east, in the complex of mountains that make up the Carpathians, there are populations in Poland, Slovakia, Romania, and the Ukraine. In the Alps, Swiss stone pine normally forms the elevational timberline, at 2,100 to 2,700 meters, where heavy snows, ice, and strong winds dominate. But centuries of heavy lumbering have reduced the extent of this subalpine forest, and the overgrazing of sheep has prevented seedlings from becoming established. As a result, this pine is missing from many places where it used to grow, and now the artificially lowered timberline is formed instead by the upper edge of the montane spruce forest. Loss of stone pine forests at high elevations has also increased the risk of avalanches on formerly stable slopes. The species is known by many names, among them *Pin cembro* or *Arolle* in French, *Arve* or *Zirbe* in German, and *Pino cembro* in Italian.

Siberian stone pine is found mainly in Russia, from the Urals east across the Central Siberian Plateau to the Aldanskoye Upland beyond Lake Baikal. There are also stands in northern Mongolia, from Lake Uvs to Uuldza. This wide-ranging pine, called *kedr* ("cedar") by the Russians, grows from moist valleys to the edge of mountain tundra high on the slopes of the Altai and the Sayans, on forty million hectares of forest land. Its nuts are an important forest product throughout this area, as a snack food and for the production of cooking oil. Nut-producing plantations have for many years been established outside villages in the Urals. A few authors still consider this pine to be merely a variety of the Swiss stone pine, but most now believe it is distinct enough to be judged a species in its own right.

The Japanese stone pine is the glutton for punishment among the *Cembrae*. It is the only soft pine found beyond latitude 70° north, and approaches within sixty miles of the Arctic Ocean, at the frosty Kolyma River delta. Its enormous range, which is exceeded among the pines only by that of the Scotch pine, extends from Khatyrka on the Bering Sea, to just south of Tokyo, and west to Lake Baikal. Its seeds are the smallest among the stone pines. It is the most unprepossessing of all pines, often dwarfed, and forming enormous areas of impenetrable scrub made denser yet by the spontaneous rooting of prostrate branches into a humus layer saturated in spring by the melting snowpack. The bleak, windswept upper slopes of many a Japanese and Siberian mountain are clothed by a mantle of this pine, called *stlannik* by the Russians and *hai-matsu* by the Japanese. A Scandinavia-sized chunk of its natural distribution area around Lake Baikal overlaps the area of Siberian stone pine, but no hybrid populations have ever been reported.

Korean stone pine grows in southeastern Siberia south along the Sikhote Alin mountains into North Korea, then westward into Manchuria. There are outlying stands in South Korea and on the Japanese islands of Honshu and Shikoku (where it is called *Chosen-matsu*). It often grows in close proximity to Japanese stone pine, but at lower elevations, and therefore in a milder climate. Its seeds are the largest among the stone pines (Figure 2.6). Long ne-

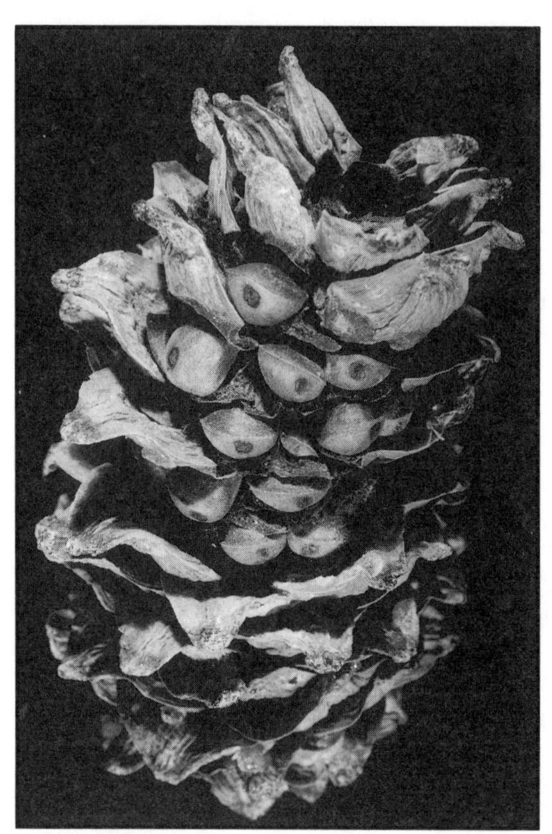

FIG. 2.6 The massive cone of Korean stone pine. Scale tips have been broken away to expose the seeds, each with its dark eye-like hilum. Utah State University Photography Service.

glected by researchers, the natural history of this vertebrate-dependent species is just now coming into focus, and is already generating controversy. It yields a valuable timber, and has therefore been subjected to heavy overcutting in its Siberian range, where the area of virgin forest shrank from seventeen million acres to half that amount between 1929 and 1983.

Whitebark pine—the focus of much of this book—is one of the true gems of our high western mountains. This short-needled, spreading-crowned tree forms open, sunny groves around the high flower-filled meadows of the northern Rockies, Sierra Nevada, and Cascades. At its northern limit in the Coast Mountains of British Columbia, it is found as low as three thousand feet above sea level. But further south, and further inland, it grows higher and higher upslope, reflecting the general increase in the elevation of the timberline. At its southern extent, far down the Sierra Nevada, it reaches heights of twelve thousand feet. It spans almost twenty degrees of latitude and 1,350 miles from north to south, exceeded among its fellow soft pines only by the Japanese stone pine. Whitebark pine enhances the high-elevation scenery of numerous American and Canadian national parks, among them Grand Teton, Yellowstone, Glacier, Waterton Lakes, Jasper, Banff, North Cascades, Olympic, Mount Rainier, Crater Lake, Lassen, Kings Canyon, Sequoia, and Yosemite. Later chapters will depict it as a keystone in the arch of these alpine ecosystems, perhaps a keystone in danger of slipping.

The five stone pines have a great deal in common besides their five-needled fascicles, wingless seeds, and non-opening cones. All are dwellers of high, cold mountain environments, although some come down into forested valleys as well. All but the Korean stone pine form the timberline somewhere in their distribution areas, and all are associated with spruces. Their faunas are strikingly similar, or they were prior to settlement. Major mammals associated with pine forests include bears, members of the deer and weasel families, squirrels, chipmunks, and voles. Several of these will be discussed later in some detail. Among the many birds common to stone pine ecosystems are jays, ravens, crows, and nutcrackers (the family Corvidae), woodpeckers, chickadees or tits, grosbeaks and crossbills, nuthatches, kinglets, and finches. Some of these too will be considered later in detail. But before a stone pine's relationship to its physical and biotic environment can be comprehended, it is necessary to know a great deal about its unique cones and the seeds they bear.

CHAPTER 3

Stone Pine Seeds and Cones

N AUGUST 22, 1805, CAPTAIN WILLIAM CLARK discovered for science the nutcracker that was later to bear his name. The Lewis and Clark expedition was then in Lemhi Pass, in the Bitterroot Mountains bordering present-day Montana and Idaho. Clark wrote in his journal that day, "I saw today a Bird of the woodpecker kind which fed on Pine burs its bill and tale white the wings black every other part of a light brown, and about the size of a robin." Nothing more was said of the pine, but the observation that the nutcracker was feeding on "burs" (cones) strongly suggests the bird was harvesting seeds of whitebark pine. Whitebarks still grow in Lemhi Pass, and several are old enough to have witnessed the captain's journey. This is the earliest recorded encounter between Clark's Nutcracker and whitebark pine.

On September 23, 1851, John Jeffrey gained the summit of a seven-thousand-foot mountain near Fort Hope, in the Fraser River country of British Columbia, and found growing on the decayed granite bedrock a group of trees he recorded as whitebark pines. Only a few trees had any cones left, and as Jeffrey explained, "Corvus Columbianus had deprived them of nearly all their seeds." This brief observation was noted in a letter to Professor J. H. Balfour from Jeffrey, a young Scottish gardener who had come to North America to continue the long-neglected plant explorations begun by David Douglas.

Three years later, J. S. Newberry, botanist of a party of United States Army topographical engineers seeking a railroad route across the Oregon Cascades, became frustrated over his inability to find intact cones of this same pine. Cones were so rare, he reported, "that, though constantly among the trees and on the lookout myself, I had for two weeks an offer, open to all of our party, of a dollar for a good cone, and no one was able to claim the reward." Yet the ground was littered with cone fragments, and Newberry was certain there had been a recent cone crop.

Newberry's observation calls attention to a characteristic unique to whitebark pine among the pines of North America: the ease with which its cones can be disassembled, and the resulting absence of whole cones lying on the forest floor (Figure 3.1). In fact, the surprising fragility of whitebark pine cones is shared by all the other stone pines but by no other pines. It deserves our attention.

In nearly all pines the time from first inception of the cone to maturation of the seeds is about twenty-five months. Typically, in July of the cone's first year, it becomes visible with the aid of a dissecting microscope as a tiny, dome-shaped mass of off-white cells within a newly forming bud. The dome

FIG. 3.1 Disarticulated whitebark pine cone scales and cores as seen in the forest litter.

grows a bit larger before and during the winter, and forms spirals of rudimentary scales, finally taking on the appearance of a miniature cone. The next spring, the "conelet" emerges from its bud looking something like a reddish or purplish pin-cushion one-fourth to one-half inch in diameter. The conelet's scales part widely, exposing two tiny ovules (the future seeds) on the upper surface of each. Airborne pollen grains strike droplets of a sugary liquid that oozes from the receptive ovules. They become trapped there, and are brought into the ovules by absorption of the liquid. The scales then close, preventing the entry of additional pollen; the conelet enlarges a bit, and then it overwinters.

Next spring the dormant pollen grains within quicken, and their sperm nuclei finally fertilize the egg cells that lie deep within the ovules. The fertilized eggs go through a complex series of developmental steps, eventually forming embryos within the ovules. By late summer these ovules develop into mature seeds. Throughout this period the cone grows rapidly, and by September it may weigh forty times what it weighed in June, thanks to the rich carbohydrate solution flowing into it from its branch.

In fact, tree physiologists believe that the carbohydrate drain places a great deal of stress on the tree, more than many trees could endure if it happened every year. Fortunately this is seldom the case. Most pines produce their cones in boom-bust cycles of irregular length. A very heavy cone year, for example, is usually followed by a bust, when cones are scarce, and then several middling years. Eventually—or so it is hypothesized—carbohydrate reserves are replenished, cone primordia form during a spell of good weather, and another booming "mast year" is in the making. Evolutionists point out the effectiveness of such cycles in outflanking predators like cone moths and seed weevils, whose populations crash in the low-crop years, rendering them incapable of wiping out the big crops that satiate them.

Actually, despite the rationalizations of evolutionary ecological theory, whitebark pine *stands* can be fairly regular cone producers even when individual *trees* are irregular. During the four most active research years in my Squaw Basin study area, there were three cone crops, one of them small, one large, and one very large. Yet studies by David Mattson of the Interagency Grizzly Bear Study Team found that, in Yellowstone National Park, individual trees produced large crops at intervals ranging from four to nine years.

In most pines, after the cone has reached maturity, it then opens and releases its seeds into the wind. In these pines, the cone opening mechanism is quite simple. Inside each cone scale are tough, coarse fibers that shrink excessively when they dry. Their shrinkage pulls the scale open and exposes the seeds within, allowing them to fall out or be blown out by the wind. The fibers give the cone structural integrity, holding it together as a unit long after it falls to the ground. Who has not at some time tried to break the scales from a pine cone? This can be a frustrating task, because those tough fibers are difficult to break. The would-be cone-disassembler must therefore flex and tug, flex and tug, until finally the last strand breaks and the scale comes off.

But whitebark pine cones and those of other stone pines behave differently. Their cone scales do not spread when they dry because they lack the coarse fibers whose shrinkage pulls other species' cone scales open. Their scales merely loosen as the cone dries, and may part just enough to reveal the seeds within, but even vigorous shaking of the cone (as might occur in a windstorm) will not dislodge the seeds. I have in my laboratory whitebark pine cones that have been stored at room temperature for twenty-five years, yet remain tightly closed, and Korean stone pine cones still closed thirteen years after picking. According to Shin-ichiro Saito, the cones of Japanese stone pine also fail to open upon maturing. I have been informed by Konstantin Krutovskii that the cones of Siberian stone pine also remain too tightly closed for their seeds to fall out. Hermann Mattes, a German scientist who has studied the mutualism in the Swiss Alps, says exactly the same about Swiss stone pine.

In 1981 I experimented with one hundred Korean stone pine cones of that year, provided me by the Kanto Forest Tree Breeding Institute at Mito, Japan. I simultaneously shook and rotated them, but succeeded in shaking loose only eleven of an estimated 8,140 seeds. Such is the effectiveness of stone pine cones in trapping their seeds inside.

Paradoxically, the closed cones of whitebark pine look very formidable. The swollen, wedge-shaped tips of their scales (the apophyses), deep purple as they approach maturity, resemble the spiky armament of prehistoric reptiles. How can so unfriendly an organ be made to yield up its seeds? The answer lies partly in the absence of the fibers discussed above, the fibers whose shrinkage pulls the cone scales open. Not only are these fibers lacking in stone pine cones, but also the scales are very thin just behind the swollen tips (Figure 3.2). Therefore they are extremely brittle, and can be easily snapped off. Further, if a scale is broken from the cone, its seeds may not even become dislodged. The most fragile part of the cone scale is at the base of the apophysis, but the seeds are placed further back on the scale. So when the scale fractures, it is the apophysis that breaks off. The seeds remain securely perched on the stubs of the scales (Figure 3.3).

Claude Crocq, a French ornithologist who has studied European nutcrackers in the Maritime Alps, watched these birds break off the scales of Swiss stone pine cones with their bills until a dozen seeds were exposed. Only then did the nutcrackers remove the seeds with quick movements of

FIG. 3.2 A sectioned mature whitebark pine cone showing prominence of the seeds and thinness of cone scales between adjacent seeds.

their bills. In 1865, Britain's famous bird biographer John Gould published a lithograph by H. C. Richter in his *Birds of Great Britain* that clearly depicted the exposure of stone pine seeds by nutcrackers (Plate III.3). In 1981 I broke off ten scale tips from each of twenty Korean stone pine cones, and rotated the cones while shaking them vigorously. Of 291 exposed seeds, only ninety-one (31 percent) were shaken loose.

So stone pine cones are like packages that can be opened to display their contents without all the contents spilling out randomly onto the ground. There is a point to the remarkable way in which these cones shelter their seeds, but to appreciate that point we must see what is so special about those seeds.

A pine seed consists of a rod-shaped embryo embedded in nutritive tissue or "endosperm," all enclosed in a hard seedcoat. It may be almost round, as in lacebark pine, or oblong, as in chilgoza pine, but it is usually egg-shaped, with a blunt end and a more pointed one. The wing, if any, is attached to the blunt end, and when the seed germinates a seedling root will emerge from the pointed end. On the blunt end, below the wing attachment, is an inconspicuous scar or depression, the hilum, where the immature seed was attached to the cone scale by a suspensor, a sort of umbilical cord through

FIG. 3.3 Whitebark pine cones that are intact (left); partially divested of scales to expose the seeds held within the core (upper right); and completely scaled showing fragility of the scale fracture zones (lower right).

which nutritive sugar solutions flowed from the branch, via the cone scale, into the developing seed. The bottom of the seed is flat, and rests in a shallow depression on the cone scale. The upper surface is usually rounded.

Pine seeds vary greatly in size and weight. The smallest are those of jack pine, which are one-eighth-inch long and number up to 250,000 per pound. The largest are those of Torrey pine, which are an inch long, and number as few as four hundred per pound. Among the soft pines wingless seeds are nearly always much larger than winged ones, which is what one would expect, given the dynamics of wind dispersal. But there are two notable exceptions: sugar pine (large and winged) and Japanese stone pine (small and wingless). Ecologically speaking, what matters most from a stone pine's point of view is that its wingless seeds exceed in size those of other conifers growing nearby. That is indeed the case with whitebark pine growing either in the Rockies or the Sierra Nevada, and with all the other stone pines as far as the available data can show (see Table 3.1). This means that, everything else being equal, the stone pine seed should be the preferred food of the nutcracker if the nutcracker can discriminate size. And of course it can.

The embryo within a seed consists of a tiny stem bearing three to eighteen needle-like seed leaves (cotyledons) packed together to form a cylinder at one end, and a rudimentary root apex at the other. It is usually light yellow in

Table 3.1 Average weight of seeds of wingless-seeded stone pines (identified in upper-case letters) and their winged-seeded coniferous associates (lower-case).

Species	Average weight (mg)
Rocky Mountains and Sierra Nevada, USA	
WHITEBARK PINE	175
Jeffrey pine	123
Red fir	71
Western white pine	17
Subalpine fir	13
Lodgepole pine	5
Mountain hemlock	4
Engelmann spruce	3
Alps, Switzerland and France	
SWISS STONE PINE	227
Silver fir	45
Mugo pine	7
European larch	6
Norway spruce	5
Siberia, Russia	
SIBERIAN STONE PINE	245
Siberian larch	10
Scotch pine	6
Korea	
KOREAN STONE PINE	554
Yezo spruce	3
Hokkaido, Japan	
JAPANESE STONE PINE	42
Maries fir	20
Sakhalin fir	10

Source: Data from USDA Forest Service (1974).

color. When the seed germinates, the enlarging embryo forces its root tip through the seedcoat downward into the soil; while its stem (the hypocotyl) elongates, and pushes its seed coat, with the cotyledons inside, above the soil surface. The brilliant, newly green cotyledons unfold like the spokes of an umbrella and throw off their empty seed coat, forming the young tree's first crown (Figure 3.4).

The moist, pearly endosperm provides the young seedling with all the nutriments of growth it will need until it can support itself on the gains of its own photosynthetic efforts. The endosperm is a storehouse of proteins, fats, and carbohydrates. These water-insoluble materials are broken down—the proteins into their component amino acids, the fats and carbohydrates into

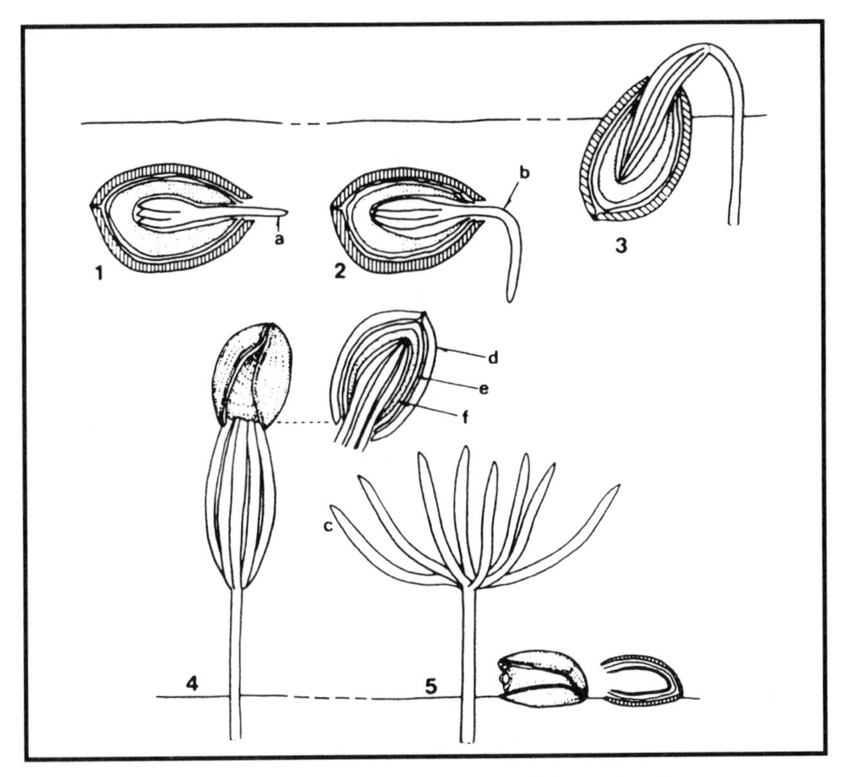

FIG. 3.4 A germinating Japanese stone pine seed. (1) The radicle (first root) penetrates the seed coat; (2) it turns downward; (3) stem growth pulls seedcoat enclosing cotyledons (seed leaves) from its subsoil cache; (4) elongating cotyledons shed empty seedcoat; (5) cotyledons spread to receive sunlight. (A) radicle; (b) embryonic stem or hypocotyl; (c) cotyledon; (d and e) outer and inner layers of seed coat; (f) endosperm. Sketches by S. Saito, courtesy of Shiretoko Museum.

soluble sugars—by the action of enzymes. Growth hormone produced in actively dividing cells of the tiny embryo acts as a messenger and traffic officer, directing these raw materials to the growth areas where they are needed, in the amounts that are required. Different pines have endosperms with distinctive proportions of nutrients. Singleleaf pinyon, for example, has a starchy seed that is about 54 percent carbohydrate, while Italian stone pine is heavy with protein (31 to 34 percent) and the *Cembrae* pines are very high in lipid content (fats and oils). Swiss stone pine kernels are about 60 percent lipids by weight; Siberian stone pine are 59 to 72 percent, Korean stone pine 52 percent, whitebark pine about 52 percent, and Japanese stone pine up to 63 percent lipids.

The proteins and lipids found in pine nuts are highly nutritious. For example, all twenty amino acids are found in the nut protein of both singleleaf and Colorado pinyon. In nuts of the latter species, seven of the nine amino acids essential to human growth are present in greater concentration than in cornmeal, a quality that made them an important food to Indians of the southwestern United States. Both species are especially rich in tryptophan and the essential sulfur-containing amino acid cystine. The essential amino acids are all present in Italian stone pine and gray pine nuts as well. The protein of whitebark pine seeds contains sixteen amino acids, with glutamic acid, lysine, and arginine especially abundant. The high lysine content is of particular interest because this amino acid is essential in maintaining a positive nitrogen balance in juncos, and might function the same way in nutcrackers.

The fatty acids that comprise the lipids of these pine nuts are also of high nutritive value. Both of the pinyon species mentioned above have fatty acids that are about 85 percent unsaturated. Italian stone pine and gray pine fatty acids are 96 and 95 percent unsaturated, respectively. The fatty acids of Korean stone pine seeds are 91 percent unsaturated, and those of whitebark pine are about 88 percent unsaturated. In all of these species the major fatty acids are oleic and linoleic acids.

Their high fat content makes these seeds an extremely rich source of calories. For example, a pound of shelled Colorado pinyon nuts provides 2,880 "large" calories, more than the food energy in that much chocolate, and almost as much as in a pound of butter. These high energy values are common among conifer seeds, wingless or not. In terms of "small" calories per gram of kernel, the reported energy values of some pines are as follows: limber pine, 7,178; lodgepole pine, 6,786; Korean stone pine, 6,953; Swiss stone pine, 6,780 to 7,742; and whitebark pine, 6,432 to 7,308. Caloric value increases as the seeds mature, as shown by the work of two of my colleagues. Harry Hutchins found that whitebark pine seeds from Squaw Basin,

Wyoming contained only 4,800 calories per gram on July 22, when they were still milky in consistency within soft seedcoats, but had increased to 7,000 calories per gram by mid-August, when firm endosperm filled their newly hardened seedcoats. Stephen Vander Wall demonstrated that limber pine seeds collected in northern Utah's Raft River Mountains doubled in energy value between late July and late August, but that the starchy seeds of single-leaf pinyon only increased theirs by about a third during the same period, and attained only 71 percent the energy value of the oilier limber pine seeds.

Even when seeds have matured, they may vary quite a bit in energy value, mainly because they vary in size or lipid content. Also, larger cones contain more seeds. Therefore the best cone, from a foraging nutcracker's viewpoint, is a large, many-seeded cone that contains large seeds of high energy value.

Whitebark pine takes its seeds very seriously indeed when it comes to distributing the tree's resources. The fraction of total energy in the seed-cone complex that is devoted to the seeds (kernels plus seed coats) is far higher in whitebark pine than in other conifers reported on. The seeds of ten whitebark pine cones I collected in Squaw Basin, Wyoming accounted for 17 to 46 percent of the total energy in the entire cone-plus-seed mass. The comparable figure for Douglas-fir is probably less than 3 percent, for ponderosa pine 12 percent, and for lodgepole pine 3 percent. That is why a sectioned whitebark pine cone reveals such a surprisingly large mass of seed material packaged in a mere jacket of swollen scale tips (Figure 3.2).

In conclusion, the seeds of stone pines are highly nutritious and packed with energy. The cones that contain them remain on the tree, unopened, giving nutcrackers the opportunity to harvest them with ease, if not without competition. How the nutcracker exploits this precious resource is the next subject of our investigation.

CHAPTER 4

The Pine Birds

IGH ON THE CONTINENTAL DIVIDE, WHERE it winds across western Wyoming, is a mosaic of subalpine meadows and forests I have visited in every month of the year. Despite the seasonal changes in the weather, the landscape, and the vegetation, one thing remains constant: I can always depend on being greeted by the guttural "kra-a-a-k" of faraway nutcrackers calling from unseen treetops. Soon one or two birds, or more, will inspect me, remind me whose mountain this is, and rapidly depart on quick-pulsed beats of white-tipped black wings. These nutcrackers are of the species *Nucifraga columbiana*, Clark's Nutcracker, native to the western United States and Canada, but I have heard Siberian Nutcrackers in a Finnish pine forest and European Nutcrackers in the French and Swiss Alps emit identical cries. This "corvid squawk," as Stephen Vander Wall has called it, one of several nutcracker calls, is not far removed from the shriller calls of jays and magpies, or the deeper ones of crows and ravens, all of which are shaped by subtle differences in size or conformation of syrinx, bronchial rings, tympanic membrane, and related anatomic determinants of voice production among members of a family.

Corvidae is the name of this, the Crow family, and in addition to the birds mentioned above the family includes choughs, rooks, jackdaws, ground jays, bush crows, and the lively Piacpiac, over a hundred species in all, distributed among twenty-six genera worldwide. Ornithologically speaking, the Corvidae belong to the order of songbirds, the Passeriformes, and are most closely related to the Shrikes and Drongos, Bowerbirds, and Birds of Paradise. They have in common ten distinct primary feathers in each wing, twelve tail feathers, tough bills, big, strong legs, and feet tipped with long claws adapted for perching, scratching, and walking.

The habitats they occupy are of a bewildering variety, from tropical forest to arctic tundra, sea level to high peaks, hot deserts to cold, timbered swamps. Their food habits are equally diverse. Many are predators or eaters of carrion. They eat fleshy fruits, seeds and grains, and even whole small plants. Many are famous for burying or otherwise hiding ("caching") their food. All of the corvids that eat pine nuts cache them by burying them in the soil. Caching food presupposes the ability to find it when it is needed, and this presupposes another quality that distinguishes the corvids—a high degree of intelligence. Indeed, they have the largest cerebral hemispheres, relative to their size, of the world's birds. Trained jackdaws and ravens have been taught to count to seven, crows to read a clock and walk a dog. But clever though such tricks may seem, they appear foolish and insignificant when compared to acts of intelligence routinely performed by corvids in the course of their everyday lives.

Among the jays are several species important to the ecology of deciduous forests. These are the European Jay, our North American Blue Jay, and the Mexican Jay. Research has shown that these corvids are vital to the establishment of oaks, whose acorns they cache in the soil to serve as a winter food supply. Acorns not recovered have a high probability of germinating. Blue Jays perform the same service with beechnuts as well.

Clark's Nutcracker and at least three other jays play a similar role in the ecology of conifer woodlands in the southwestern United States and Mexico. These woodlands, made up of pinyon pines and junipers, occupy about forty-seven million acres in the southwest and an enormous area in Mexico (Plate III.1, 4). They form the major vegetation type between low-elevation desert scrub and the denser forests of tall trees further upslope. The species of pinyon pine share several traits: they are low in stature, bushy-crowned, short-needled, highly resinous, very drought-resistant, and slow-growing. All have large, wingless seeds that are deeply recessed into the upper surface of the cone scales, which are held there for some time by membranous portions of cone-scale tissue. The large, nutritious seeds are enclosed within coats or shells, varying among the species from one-tenth to eight-tenths of a millimeter in thickness, and are highly attractive to birds and mammals.

The three jays and single nutcracker species that forage for pine nuts in these woodlands do so in varied ways, and with different degrees of necessity. The Scrub Jay consists of fifteen subspecies spread from Florida, where it has been called the "Florida Jay," to California, home of the "California Jay." This bird, which is aggressively territorial in the West, has an azure-blue crown, rump, tail, and wings, with grayish underparts and a brownish back. It is about twelve inches long and weighs about three ounces. It inhabits a wide variety of scrub and woodland communities, and its food intake reflects that variety: it eats seeds and nuts, including acorns; insects, berries, and fleshy fruits; young birds and eggs of other species; and even small mammals.

Tony Angell has described the Southern California suburbs as a paradise for Scrub Jays that "prowl the school yards for discarded lunches and the edges of parking lots for whatever fare can be found." Pine nuts are believed to form less than 60 percent of the Scrub Jay winter diet, and it is not likely they are fed in large numbers to nestlings. The Scrub Jay forages individually or in small groups seldom exceeding five birds, and a single bird may store six thousand pine nuts in the fall.

Scrub Jays are not especially well-equipped to deal with pine nuts and the cones that bear them. Their relatively short (three-quarter-inch), blunt bills cannot tear apart closed cones, so the birds can harvest seeds only from cones already opened. Early in the season they are therefore reduced to the status of kleptoparasites, startling foraging nutcrackers and Pinyon Jays with aggressive movements in hope of making them drop seeds which the Scrub Jay can then make off with. The Scrub Jay has no structural adaptations allowing it to carry a large number of seeds. Four or five pinyon nuts are apparently all its mouth and bill can hold. These seeds are usually transported less than 150 feet before being cached, though occasionally this shortest-winged of the pine nut-caching corvids will weakly flap and glide for up to one-third of a mile before caching. When it gets to its destination, it makes caches which almost always contain only a single seed. Its short-distance flights make the Scrub Jay an unlikely agent for establishing new woodlands, but Russell Balda has suggested that this limitation may enhance its ability to regenerate pinyon pines within the woodlands where the seeds are harvested.

Steller's Jay seems a bit more at home with pine nuts. Florence Merriam Bailey neatly summed up this bird as "the mountain king of the jays, with his white-streaked, high black crest, his handsome dark blue wings and tail, and turquoise rump." Its bill is no more useful than that of the Scrub Jay for hammering on cones, but at least Steller's Jay has an esophagus that can be stretched to accommodate about eighteen pinyon nuts, and it has been known to fly with them nearly two miles to its caching area. Steller's Jay is also larger and heavier than the Scrub Jay. Its caches, like those of Scrub Jays, are usually single seeds, occasionally a pair or a trio. Steller's Jays nest and forage in a wide variety of wooded and forested habitats throughout western North America, so like Scrub Jays, they are not heavily dependent on stored pine nuts. Even so, a single bird may cache as many as 3,900 pinyon nuts in sites where there is a good chance of germination. The tendency of Steller's Jay to cache seeds in forested areas above the woodland may help move the pinyon pine upslope at times when a changing climate permits it to become established there.

The low level of dependence on pine nuts in both Scrub and Steller's Jays may be related to their nesting habits: both species lay their eggs in April and May, and when the nestlings hatch there is a wide variety of foods that can be fed them. Their impact on the ecology of the pinyon-juniper woodland is less obvious than that of the Pinyon Jay because there are not very many of them, they do not cache very many seeds, and they do not transport seeds a great distance.

In all these respects, the Pinyon Jay commands attention. The Pinyon Jay is slightly heavier and longer-winged than the Steller's Jay, though a bit shorter. Its plumage is lighter than that of Steller's Jay, varying in hue from deep cyanine through cobalt, azure, and flax-flower blue. The coloration results not from pigmentation, but, as in the other blue jays of North America, from the play of light in the structure of the feather barbs. The Pinyon Jay is far more deeply involved in the biology of the woodland, more dependent upon and adapted to the nuts of the pinyon pine (Plate 111.2).

These year-round residents of the woodland and adjacent forest breed in colonies that may contain many nests. Noisy flocks of fifty to more than four hundred and occasionally thousands of birds start to forage for ripening pinyon nuts in late August. Their long, strong bills are used to peck the cone from its limb, hammer it until the scales are shredded, and deftly remove the intact seeds. The effective length of the Pinyon Jay's bill is increased by not having bristles around the nostrils, allowing the bird to probe deeply into resinous cones without becoming fouled by pitch. The disadvantage of this

Table 4.1 Vital statistics of the North American corvids that cache pine nuts

Species	Length (inches)	Weight (ounces)	Bill length (inches)	Maximum load (ounces)	Maximum speed (mph)	Maximum flight (miles)
Scrub Jay	12	3	.75	.05	18	.33
Steller's Jay	12	4	.75	.20	22	2
Pinyon Jay	11	4	1.00	.60	26	6
Clark's Nutcracker	12	5	1.25	1.10	29	18

Source: Data from various sources.
Values are converted from metric units, therefore they are approximate.

adaptation is that these bristles normally function in other birds to reduce heat loss from the nostrils. Roosting jays compensate for this loss by snugly tucking their bills in their scapulars on cold nights.

When cones mature and part their scales, Pinyon Jays quickly harvest the exposed nuts. They distinguish visually between sound, chocolate-colored seeds and buff-colored empty ones, rejecting the latter and leaving them in the cone. Apparently a vital embryo and endosperm are necessary for a pine seed coat to develop the dark pigmentation and waxy surface that typify the filled, healthy seed. Pinyon Jays select these seeds, but test them further by "clicking" them several times in the bill, and holding them in the bill for a second or longer, apparently to assess their soundness and weight. Experiments show that while "naive" Pinyon Jays—young birds reared in captivity on dog food and baby mice—are strongly attracted to pinyon seeds, they must learn to discriminate between good and bad seeds. A seed that is to be eaten is held between the feet and a perch, and is vigorously pounded with the bill until the seed coat cracks open. The jay then deftly removes and swallows the kernel. The Pinyon Jay's esophagus becomes stretchable when its muscles are relaxed, providing space for as many as fifty pinyon nuts. Once "loaded" the jays fly with strong wingbeats through or above the woodland canopy to communal cache areas as far as seven miles distant, but usually between one-half and three miles.

Pinyon Jays are famous for their prodigious flight abilities. Charles Aiken observed at Fort Garland, New Mexico in October, 1874:

> Probably a hundred of these birds in a dense rounded mass, performing evolutions high in the air . . . sweeping in wide circles, shooting straight ahead, and wildly diving and whirling about, in precisely the same manner that our common pigeons do when pursued by a hawk. (Bailey 1928, p. 500–501)

This "singular performance" continued for two hours, despite the absence of a hawk to keep the jays properly stirred up.

At the cache site, which may be in woodland, in lower-elevation grassland, or in higher-elevation forest, seeds are held in the bill one at a time, and inserted an inch or so into soil previously loosened with one or two jabs of the bill. Most Pinyon Jay caches consist of one seed, but nearly half the seeds cached may be in groups as large as six or seven. A flock of two hundred fifty Pinyon Jays in New Mexico was estimated by David Ligon to cache thirty thousand seeds daily, with a seasonal total of four and one-half million. A single bird caches as many as twenty thousand seeds in a season. These food stores are of great importance to the Pinyon Jay, and comprise up to 90 percent of its diet from November through February. Ample stored pine nuts provide the Pinyon Jay the energy it needs for gonads development, courtship, nest-building, egg-laying, and incubation of the eggs. But in years when the pinyon crop fails, Pinyon Jays erupt (or "irrupt") from their breeding grounds, wandering widely in search of food. Although Pinyon Jay nestlings live on a diet of mostly insects, pinyon nuts are important as a reserve.

Stephen Vander Wall and Russell Balda have described these three species of jays as a graded series showing increasing specialization to a natural economy based on the seed of the pinyon pine, from the generalist Scrub Jay to the highly specialized Pinyon Jay. Thus, going up the scale—which is just a device for viewing diversity of adaptations, and does not imply phylogeny, or line of descent—we encounter larger, stronger, bigger-billed, further-flying birds that have increasingly specialized seed-carrying equipment, that can harvest more seeds and make more and larger caches, and that in turn become more dependent on the pines and more important as dispersers of their seeds. The series culminates in the largest, strongest, biggest-billed, furthest-flying and best-equipped of all corvids that feed on conifer seeds, and presumably the most advanced in its evolution—the nutcracker.

CHAPTER 5

The Top-of-the-Line Pinivore

N JULY 13, 1979, HARRY HUTCHINS SAW BOTH the beginning and the end of Clark's Nutcracker's period of dependency on pine nuts. Harry had recently come west from Michigan, with bachelor's degrees in both forestry and wildlife science, to study pines and birds for a master's degree at Utah State University. Having become acclimated to the altitude and to Rocky Mountain hailstorms, he found himself 9,700 feet above sea level at Surprise Lake, Grand Teton National Park, binoculars hanging from his neck and notebook in hand. Harry was getting to know Clark's Nutcracker. Among his many observations that day, two have contextual significance: he saw a nutcracker foraging for unripe seeds from the 1979 crop in an immature, purple whitebark pine cone; and he observed a nutcracker parent digging up cached nuts from the 1978 crop, which it then fed to a nagging youngster.

Clark's Nutcracker is the only nutcracker native to North America. In size and shape it is a slightly enhanced Pinyon Jay: its body weight is nearly five ounces versus about four; it is about twelve-and-a-half inches long versus eleven; its wingspread is an impressive twenty-two inches versus eighteen; and its stout, black, chisel-like bill measures one and one-quarter inches from tip to nostrils, against the Pinyon Jay's inch. Its plumage—a smoky gray head, neck, and body, black, white-tipped wings, and black-centered tail with white edges—gives it a dressed-up look, something like an understated penguin (Figure 5.1). It can attain an age of twelve years.

FIG. 5.1 Clark's Nutcracker, *Nucifraga columbiana*. Drawing by Louis Agassiz Fuertes.

The only other member of the genus is the Eurasian Nutcracker, *Nucifraga caryocatactes*, which is distinguished from other corvids by its dark chocolate brown to pale brown plumage with pear-shaped white spots at the tip of each feather. There are eight to ten subspecies of this broadly-distributed bird, depending on the taxonomist consulted. Those discussed in this book will be referred to as the European, Siberian, and Japanese Nutcrackers instead of subspecies *caryocatactes*, *macrorhynchos*, and *japonica* (or *japonicus*) respectively.

Though Clark's Nutcracker comes down into the Southwestern pinyon-juniper woodlands from higher elevations and plays an important biological role there as an eater and cacher of pine nuts, its real home is in the upper-montane forests of the Rocky Mountains, Cascades, and Sierra Nevada. Florence Merriam Bailey has contrasted the behavior of the relatively timid Gray or Canada Jays, which "retreat to the protecting depths of the dense Hudsonian forest," to that of the bold nutcrackers, which "fly up into the open, among the wind-beaten dwarfs of timberline." It is from those heights, up to 12,600 feet on Pecos Baldy (New Mexico) by Bailey's observation, that nutcrackers descend in what James Neilsen called their wonderful plunge flight, typified by a thousand-foot nose-dive to a valley floor, broken by an upward swoop causing an audible roar of wind in wings. In high mountains it builds its well-concealed nest, a tightly constructed mass of woody twigs lined with grass and shredded bark, placed high in a conifer. The nest must be well-insulated, because Clark's Nutcracker is one of North America's earliest-breeding passerines, laying two to four brown-spotted pale green eggs as early as late February. After sixteen to seventeen days of incubation, the eggs hatch. At that time the nutcracker's world is a snowscape, and there is almost no food available except for the pine nuts cached in the soil the previous fall. When summer comes the nutcracker will feed on a variety of insects, even hawking moths like a flycatcher. It will raid birds' nests to eat the eggs and nestlings, kill an occasional vole or ground squirrel, and even raid a trapline for meat baits. But during the long fall, winter, and spring, survival depends on cached food, and nestlings are fed shelled pine nuts almost exclusively.

A nutcracker's dependence on pine nuts extends beyond the nestling stage. For surprisingly long after it has fledged (grown its flight feathers), a young nutcracker must still be fed by its parents. Vander Wall and Hutchins have graphically described the behavior of adults and young at this time:

> While adult nutcrackers searched for cached seeds, groups of young usually waited in nearby trees. Young frequently gave *kra-a-a* calls similar to those given by adults but higher pitched and repeated more frequently (20–30 times/min). When adults approached with small loads of shelled seeds in their pouches, the young flew or hopped toward the adults and began giving the hunger call, a frantic *near*. . . . As a young approached an adult its wings were held partially outstretched and quivered rapidly. The mouth was gaped wide and the head tilted slightly upward. The adult then ejected seeds from its pouch with quick jerks of the head and, holding the seeds one at a time in the tip of the bill, inserted them deep into the young's mouth. Each young typically received 3–7 whole seeds in quick succession. We never saw adults feed young anything other than Whitebark Pine seeds. (Vander Wall and Hutchins 1983, p. 97)

On other occasions, the young hopped along behind the adult, nagging it with hunger calls, while the adult removed seeds from the cache. The seeds were shelled and immediately fed to the young.

Postfledging dependency continues for thirteen to fourteen weeks, until about mid-July. By then significant numbers of cached seeds are germinating, their elongating seedling stems forcing the attached seed coats through the surface of the soil. This gives the young nutcrackers their first opportunity to forage independently of the adults, and they sever the seed coats from the stems to recover whatever endosperm remains within. Now the new seed crop is becoming attractive, and the young begin to learn cone foraging by watching the adults. Since survival depends on harvesting and storing seeds, it is imperative that the adults be free of parental responsibilities when harvest time arrives, even if juveniles come close to starving while developing their foraging technique by trial and error. Mervin Giuntoli and L. Richard Mewaldt report that in 281 stomachs from nutcrackers collected in Montana between November and April in the late 1940s, conifer seed comprised 70 to 100 percent of the contents.

Adult nutcrackers are highly effective seed harvesters and cachers. Their powerful bills are more than a match for the cones they attack. Their bills can riddle and shred hard ripened cones like those of the limber pine (Figure 5.2), and peck cones from the branch; but they can also makes slits as delicate as those of a surgeon's scalpel on the soft green scales of Douglas-fir cones, when the nutcracker desires to eat seeds of that species. Juveniles are much less effective at foraging, often losing their footing on swaying

FIG. 5.2 A limber pine cone "attacked" by Clark's Nutcracker. The scales have been pecked and shredded to facilitate removal of the seeds.

branches and missing the targets of their bill stabs. But they improve during the season. Clark's Nutcracker uses the same visual cues as Pinyon Jays in avoiding empty seeds, and also uses bill-weighing and bill-clicking. It opens seeds by pounding them with its bill against an "anvil," as does the Pinyon Jay, but it is also capable of crushing them in the bill, then removing the kernel with the tongue and swallowing it. The Eurasian Nutcracker has a bump on its lower mandible which is believed to facilitate the cracking of nutshells. Shell fragments are commonly ingested, presumably to be used for grinding food in the nutcracker's muscular stomach. But seeds intended for cache-storage are neither shelled nor passed on to the stomach. Instead, they are diverted into the structure that sets nutcrackers apart from all other birds, and that exquisitely adapts them to life among the pines: the sublingual pouch.

Technically a diverticulum, or sacklike extension, of the floor of the mouth, the sublingual ("under the tongue") pouch is carry-on luggage for birds traveling with pine nuts. Each nut to be pouched is brought into the oral cavity, and dropped into the pouch through an opening at the base of the tongue. The pouch wall is thin, wrinkled, and elastic, and stretches as seeds are added, swelling almost to the size of a walnut when fully packed (Figure 5.3). The capacity of a pouch stuffed with twenty-eight singleleaf

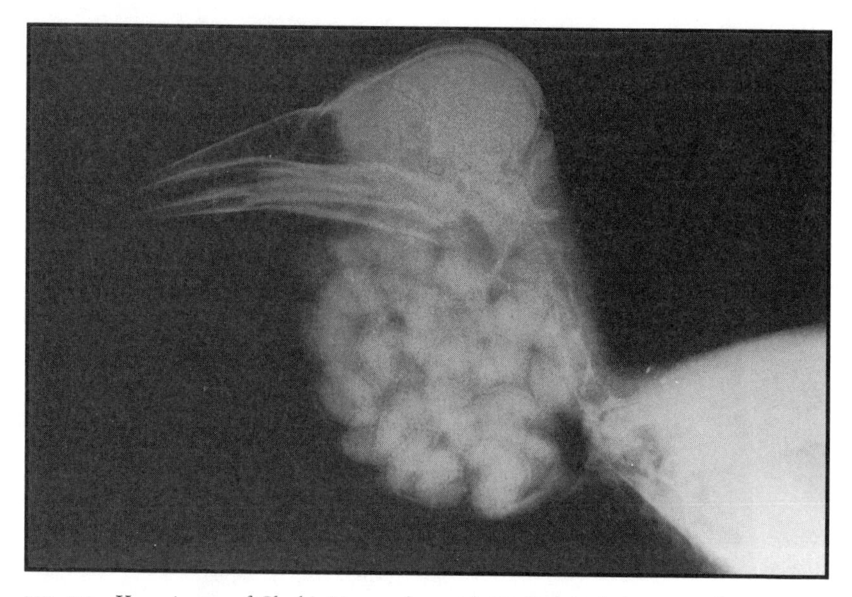

FIG. 5.3 X-ray image of Clark's Nutcracker with 28 singleleaf pinyon seeds in its sublingual pouch. The seeds weigh 31 grams; the bird weighs 141 grams. These seeds are about twice the size of whitebark pine seeds. Photo courtesy of S. B. Vander Wall.

pinyon nuts is about 28.5 milliliters, sufficiently capacious for ninety seeds of Colorado pinyon. A Clark's Nutcracker sacrificed for science a century ago in Montana had eighty-two whitebark pine seeds in its pouch. As many as 134 Swiss stone pine seeds, 167 Siberian stone pine seeds, and up to 218 of the smaller seeds of Japanese stone pine have been found in the pouches of Eurasian Nutcrackers.

The act of accumulating a full pouch of pine nuts hardly seems like the frenzy of stereotyped movements some behaviorists expect of birds. Indeed, the nutcracker is a very discriminating bird, making numerous choices throughout the foraging process. It is choosy about the species it forages on, and, as Vander Wall has shown, switches from one to another when doing so enhances its foraging efficiency. But it does not find all "prey" trees of the same species equally attractive. Given a choice, nutcrackers prefer pine populations with the largest cone crops, and within these populations they show preference for the trees that bear the most cones. Kerry Christensen, Thomas Whitham, and Russell Balda noted these and other behavioral quirks of Clark's Nutcrackers in field and laboratory studies in northern Arizona. They found that the more cones a pinyon pine had, the more would be emptied by nutcrackers and the larger proportion of its seeds would be taken. Nutcrackers also preferred larger cones with more seeds, and cones with a higher proportion of viable seeds. Add to these choices that of selecting individual seeds that pass the color and bill-clicking tests, and what emerges is the image of a very careful shopper, what ecologists refer to as an optimal forager. As we will see, such behavior may well have influenced the evolution of the pines that feed nutcrackers.

Even having stowed a full pouch of pine nuts, Clark's Nutcracker remains a prodigious flyer. A fully burdened bird flies up with strong, rhythmic wingbeats, high above the forest canopy, heading directly for its caching area. On sunny days when thermals are rising, they will often ride these to gain elevation. Reaching speeds of thirty miles per hour, and carrying a load of seeds that can exceed 20 percent of its body weight, a nutcracker will fly several miles to a caching area that may be at a significantly higher elevation. In Arizona's San Francisco Peaks nutcrackers were frequently seen caching seeds six to nine miles from their harvest site and nearly two thousand feet above it. The maximum distance between the harvest and caching areas, observed by Balda in the San Francisco Peaks, is an impressive eighteen miles. In the Sierra Nevada, Diana Tomback observed flights of seven and a half miles. Hermann Mattes tracked European Nutcrackers in Switzerland's Upper Engadine Valley nine miles, to caching areas 2,300 feet above the seed harvest site.

Nutcrackers seem somewhat inconsistent in choosing places to make their caches. All the pinyon caching done by the Arizona nutcrackers observed by

Stephen Vander Wall and Russell Balda occurred on steep slopes that faced to the south. One of these slopes, which was about six acres in extent, appeared to be a communal area where all members of the flock cached their seeds. In the Sierra Nevada, Diana Tomback found a more complicated situation among nutcrackers caching whitebark, singleleaf pinyon, and Jeffrey pine seeds. As in Arizona, there were communal caching sites on southern exposures. Some nutcrackers flew more than a mile to cache at these locations, but most landed on the ground not far from the trees from which they had taken their pouch-loads of seeds, and proceeded to make a series of caches. They wandered about the forest, seemingly at random, scattering their caches in varied terrain. In both studies the investigators interpreted southern-exposure caching as a means of storing seeds where snow would be less deep and would melt out earlier in the spring. They also stressed the advantages of dry storage to better preserve the seeds.

In a study conducted in Squaw Basin, Wyoming, Harry Hutchins found that seeds were cached either within one hundred meters of the trees from which they were harvested, or over two miles away on ledges of the steep, southwest-facing Breccia Cliffs. The locally cached seeds were buried indiscriminately in deep shade, on open meadow, on a streambank—even in a puddle of water!—and on all exposures. The nutcrackers, caching singly or in a flock that sometimes reached 150 birds, had been frantically removing as many seeds as possible from the treetops, hiding them safely in the soil, seemingly concerned mainly with harvesting them before the Red Squirrels did. But by mid-October, with the crop considerably diminished, they started digging up their Squaw Basin caches and flying the seeds to the Breccia Cliffs to be re-cached. The Russian ecologist A. A. Mezhennyi (1964) once observed a similar occurrence in Yakutia, in east-central Siberia. He reported that Siberian Nutcrackers recached *kedr* seeds ten miles from the trees they came from, after initially concealing them in the nearby forest.

Having selected an area in which to cache its seeds, a nutcracker will usually perch in a tree, cry loudly several times ("*kra-a-a*"), look around with furtive glances, and fly to the ground. If a Steller's Jay makes an appearance, the nutcracker, aware of the jay's thieving ways, will fly off or wait for the jay to leave. The bill is pushed an inch or so into the ground and quickly worked from side to side, loosening the soil. Seeds are brought singly from the pouch into the bill with jerky head movements, held briefly in the tip of the bill, and quickly pushed about an inch into the loose soil. Harry Hutchins dug up 157 caches of whitebark pine seeds and found them to contain from one to fourteen seeds, with an average of about three. Thirty-five percent of the caches consisted of only one seed, 18 percent were two-seeded, and 18 percent three-seeded. Curt Dimmick has reported a cache of twenty-four whitebark pine seeds. Hermann Mattes found in the Swiss Alps that European Nutcrackers made caches of up to twenty-four *Zirbe* nuts, with the average about three to four nuts. The Siberian Nutcracker makes caches of up to sixty nuts of *kedr*, and the Japanese Nutcracker up to twenty-three nuts of the *haimatsu.*

After burying their seeds, nutcrackers often conceal a cache site by covering it with soil, a few small stones, a pine cone, or whatever presents itself, even a lump of moose dung. These caches are often made at the base of a tree, shrub, or rock, beneath the edge of a moss pad, or in forest-floor litter. The Eurasian Nutcracker appears to be more inclined than Clark's Nutcracker to cache seeds in decaying logs and stumps, or beneath the lichens or bark of tree-trunks and limbs, and both species avoid caching in clumps of grass or sedge.

In 1980, a year of whitebark pine cone abundance in Squaw Basin, Wyoming, this show ended about the first of November. By then virtually no cones were left on the trees except for seedless remnants that had been riddled by nutcrackers before the seeds could mature. Cones containing mature seeds had all been cut by squirrels or fragmented by nutcrackers. An enormous number of seeds lay just beneath the soil surface, in many thousands of small caches throughout the forest, on the adjacent meadows, and high on the dominating Breccia Cliffs. How many? Russian scientists have made estimates over the years of the magnitude of nutcracker caching activity. D. I. Bibikov, for example, calculated in 1948 that between 1,600 and 13,600 Siberian stone pine seeds lay beneath every upland acre of boreal forest in Siberia. N. F. Reimers put the stored crop at seventeen thousand seeds per acre, averaging nearly five pounds of seeds. Vander Wall and Balda made high and low estimates of the quantity of pinyon seeds cached by a population of 150 Clark's Nutcrackers in northern Arizona in 1969. The average of these estimates is over four million seeds weighing almost nineteen hundred pounds, of which 650 pounds was fat!

How many seeds does a nutcracker hide in the ground? The available estimates are fraught with uncertainty, because there are so many variables to be considered, and it is difficult to make precise field observations; but they do give us a picture of the magnitude of the seed-caching enterprise that absorbs a nutcracker's energies.

Vander Wall and Balda assumed, based on intensive observations in northern Arizona, that a Clark's Nutcracker made four to six trips daily for one hundred days, carrying an average of fifty-five pinyon pine nuts per trip. Therefore a single nutcracker would have cached twenty-two thousand to thirty-three thousand seeds. When they calculated the caloric value of these seeds, and compared it with the calories that a nutcracker would need to maintain itself metabolically between mid-October and mid-April, they found that a nutcracker's caches contained between 2.2 and 3.3 times the energy necessary for survival.

Making somewhat different assumptions about Clark's Nutcrackers in her Sierra Nevada study area, Tomback calculated a potential storage of thirty-five thousand whitebark pine seeds placed in ninety-five hundred caches, which was, according to her calculations, three to five times as many seeds as a nutcracker needed to survive the winter and feed its young. Curt Dimmick estimated that adults in his Tioga Pass study area cached almost eighty-nine thousand seeds in 1990, while juvenile birds cached far less, about thirty-four thousand each.

Hutchins estimated the number of whitebark pine seeds a Clark's Nutcracker would cache on the Breccia Cliffs after harvesting them in Squaw Basin, a 4.2-mile round trip. He started from such basic field data as the time it takes to fill a pouch with seeds, number of seeds per filled pouch, flight time, time required to cache a pouchload of seeds, nutcracker time spent preening and maintaining itself, and number of hours per day of activity. According to his calculations, a nutcracker making 1,053 trips in an eighty-day period would cache 98,000 seeds. At 3.2 seeds per cache, that comes out to about 30,600 caches.

Hermann Mattes conducted a similar exercise for the European Nutcracker in a seventy-acre Swiss forest that was home to five breeding pairs. When Swiss stone pine produced a medium-sized seed crop in 1974, each bird stored 109,000 seeds; but the following year, when the crop was a scant one, only 47,000 nuts were stored. Mattes found an average cache size to be three to four nuts, so the number of caches made in a medium seed year could be as high as 36,300. Even after the modest seed crop of 1975, Mattes found that the nutcrackers of the Staz forest used only 82 percent of the seed they had cached in the forest floor. His results seem to confirm what Hutchins and I suspected from the Squaw Basin study—that nutcrackers will continue to harvest and cache pine nuts until there are no more left in the trees, regardless of how far these exceed their needs.

Russell Balda intimated as much in a symposium at the Seventeenth International Ornithological Congress in Berlin:

> One is always impressed with the industriousness of (nutcrackers) and the great effort birds will go through to obtain these prized seeds. The fixation on pine seeds seems unparalleled in the bird world . . . it is the motivation or drive that results in the caching of many more seeds than the bird can possibly use. (Balda 1978)

So powerful a drive, coupled with what appears to be a lifelong and year-round dependence on pine nuts, raises a dilemma for nutcrackers. On the one hand, if a crop of pine seeds sufficient to the birds' needs indeed materializes, and is duly concealed in the soil, how can the nutcrackers ever find them again when they are needed? And on the other hand, if the year comes that no such crop is produced, what then do these feathered pinivores do to stave off extinction? Both questions are central to understanding the mutualistic relationship of nutcrackers and pines. Both require an answer.

CHAPTER 6

Memories

OW DO NUTCRACKERS FIND THEIR CACHES? Over the years many naturalists have been intrigued by the unhesitating way in which birds of both nutcracker species fly to the ground and unearth pine nuts with a few swipes of the bill, as if knowing exactly where to dig. Even snow is no deterrent. Claude Crocq has described how European Nutcrackers in the French Alps recover pine nuts by tunneling to caches concealed by more than a meter of snow. Their tunnels frequently curve beneath the snow, and while excavating, the completely submerged bird cannot see out (Figure 6.1). Yet more than three-fourths of 125 such *galeries* bore the tell-tale evidence of seed-coat fragments, attesting to the excavator's success.

Vander Wall and Hutchins watched a Clark's Nutcracker peck through a half-inch crust of ice at the edge of a melting snowpack to recover whitebark pine seeds buried beneath the ice. The bird spent seventy-five seconds on this task, no trivial expenditure of its time. What makes such feats remarkable is that winter snows so completely modify the cache environment that a bird must find many caches in order to survive, and that some caches may be retrieved ten months after having been made.

Do nutcrackers smell the buried seeds? Pine seeds are often sticky with drops of resin that have gotten on them while in the cone. Conceivably, the resinous odor could be detected by a nutcracker even through the soil, especially if the cache is a large one. True, most corvids do not have a highly developed sense of smell, but Black-billed Magpies have been known to detect a decaying chicken hidden beneath an underwater rock, and have been at-

FIG. 6.1 This nutcracker made its seed cache close to the base of a tree, which serves as a visual landmark. But to recover the seeds it must start its tunnel at distance a from the tree if the snow is shallow (c); or at distance b if the snow is deep (d). Sketch by Claude Crocq.

tracted to sub-soil caches of raisins and suet that were made odoriferous with cod liver oil. Turkey Vultures use scent to locate carrion; storm-petrels, shearwaters, fulmars, and albatrosses use scent to find fish beneath the surface of the sea; and pigeons use scent to help navigate.

The New Mexico ornithologist J. Stokley Ligon once puzzled over the possibility of nutcrackers using scent to locate another kind of food:

> The skill with which they locate a source of food is astonishing. On several occasions the author has hung the carcass of a deer in the shade of some thick conifer, returning later to find two or three Clark's Crows helping themselves to a meal from the exposed flesh where the animal has been drawn. Although it is contrary to research findings, this ability to spot a feast so quickly indicates that such food might be located by scent. (J. S. Ligon 1961, pp. 205–206)

As Ligon points out, his observations can also be explained by other means—attraction to rifle fire, for example—but it seems clear that any serious attempt to explain the cache-discovery behavior of nutcrackers must confront the possibility of scent. Another possibility is that nutcrackers can visually detect the soil disturbance resulting from seed caching. Or maybe these birds search for caches *at random*—land anywhere, poke the bill around, and get lucky often enough to come up with sufficient food to survive on. Implausible perhaps, but surely a possibility.

A better system would be a *directed* random search, similar to the above, but restricted to particular kinds of places where nutcrackers habitually make caches. A random search directed at likely communal caching areas on south-facing slopes, for example, or on windswept ridgetops, might work because caches are highly concentrated in such places.

A final possibility is that *spatial memory* is at work in cache recovery, that a nutcracker *remembers* in fine detail where it has placed each of its caches. The use of memory would require the bird to relocate the cache site by using visual cues; in other words, it would remember the location of a cache in relation to other objects in the landscape.

Field observations have tended to narrow on memory as the effective mechanism. These observations have almost all been similar to those of Crocq, made in the snow where nutcrackers have left visible evidence in the form of tunnels in the snow, footprints, bill marks, and cracked nutshells. The method has been employed in Europe, Siberia, and California. The underlying idea behind these observations is that snow cover precludes the use of a sense of smell, and conceals any scarring of the ground surface. If the "success rate" (number of caches found compared with the number of attempts) is high enough, then random and directed random searches can also be eliminated, leaving memory as the only alternative.

But these after-the-fact divinations of bill marks in the snow cannot really settle the question of how nutcrackers find cached seeds, because the method is flawed. It incorrectly assumes that the absence of cracked shells at a potential cache site means no seeds were cached there and that the bird guessed wrongly about the location of a cache. Actually, a nutcracker may have found seeds there, but pouched them in order to eat them elsewhere. Or rodents may have raided the cache, emptying it and causing the nutcracker to come up empty-billed. More seriously, there are no controls over the behavior of wild birds in the natural environment, and therefore no way of telling whether a bird is indeed relocating one of its own caches. Consequently there is no way of knowing whether memory is involved. Such field observations under uncontrolled conditions are valuable for generating testable hypotheses, but cannot answer the questions those hypotheses raise.

The memory hypothesis received strong support from an experiment performed by Krushinskaya (1970) in the Soviet Union. Krushinskaya's point of departure was the assumption that spatial perceptions, including spatial memory, are stored and processed in a part of the forebrain known as the hippocampus. She reasoned that Eurasian Nutcrackers from which the hippocampus had been removed would not fare well in relocating their own seed caches made in a small aviary. She was correct. Nutcrackers lacking a hippocampus succeeded in finding only 13 percent of their stored seeds, while birds of the control groups located 78 to 91 percent of their caches. This was strong presumptive evidence that nutcrackers remember where their caches are, but it did not provide the mechanism with which they find the caches. What memories exactly do nutcrackers store in their brains? This question seems finally to have been answered by a set of elegant experiments performed by Stephen Vander Wall.

Vander Wall erected five hypotheses, each aimed at one of the five possible explanations of cache relocation mentioned above. Then he derived two to four predictions that followed from each hypothesis. For example, Hypothesis 1 stated that nutcrackers find cached seeds by olfactory cues (scent) emanating from the cache. If this is true, one can predict that birds should find each other's caches and those made by the experimenter as well as its own, and the birds should not dig for a cache from which the seeds have been removed. Vander Wall performed four experiments in an open-air aviary about thirty feet square in the Green Canyon Ecology Center compound maintained by Utah State University near Logan.

The aviary was built largely of scrap lumber, and cost Vander Wall only about thirty dollars cash. But getting to that point had been a long and frustrating journey. Vander Wall had originally obtained permits from the U.S. Fish and Wildlife Service and the State of Utah allowing him to capture nutcrackers in 1978. He had gone through the tedious procedure of setting up black nylon mist nets (strips of mesh about six feet high and thirty feet long) between cone-laden singleleaf pinyon trees in the Raft River Mountains over a hundred miles to the west. Eventually he netted several birds and brought them to Logan. It was during the time of fall nut harvest for these birds, so Vander Wall had to collect a large quantity of pine nuts with which to feed them, as well as to use in his planned experiments. The large cage in which the nutcrackers were sequestered was adjacent to a goat pen where behavioral studies were in progress. True to their reputation, one of the rams, provoked by unknown stimuli, butted down the common wall and liberated Vander Wall's nutcrackers. By then the nut harvest in the mountains was completed, and it was no longer possible to catch more of the elusive corvids. It was two years later, in 1980, that Vander Wall was finally able to reestablish his experimental setup, capture new birds, and test his hypotheses.

Within this enclosure, Vander Wall created a miniature nutcracker world. The finely raked soil and gravel floor (to four inches deep) was underlain with wire mesh to prevent rodents from tunneling in from outside and stealing pine nuts. Sixty-nine logs, shrubs, large rocks, and stumps were placed at the intersections of a grid (Figure 6.2). From his elevated booth, Vander Wall could observe where caches were made or recovered, and by which bird, and could precisely map the locations. He used four captive Clark's Nutcrackers, named Orange, Red, Blue and Green after their leg-bands. Orange and Red could be induced to cache by providing them with seeds, or to dig up caches by depriving them of seeds and making them hungry. Blue and Green were for some reason disinclined to make their own caches.

To test the olfaction hypothesis, Vander Wall allowed Orange and Red to cache singleleaf pinyon seeds privately for 441 and 523 minutes respectively, spread over several visits. Though Orange and Red were not allowed to watch each other cache, both were observed some of the time by Blue and Green, who did not cache. When Red and Orange had finished caching, Vander Wall

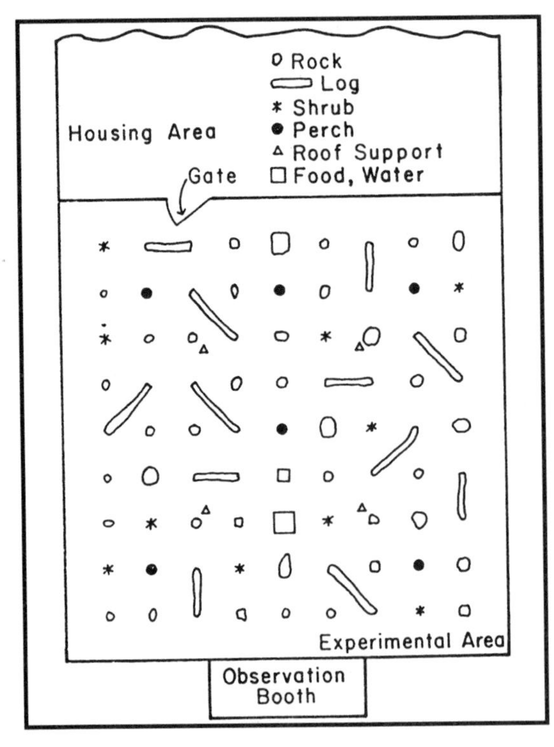

FIG. 6.2 Layout of enclosure used by Vander Wall to test hypotheses of seed-cache recovery by Clark's Nutcracker. Reprinted with permission from *Animal Behaviour*.

carefully removed twenty-five of each bird's caches, and made 102 artificial caches of his own. On two days following the bouts of caching, Blue and Green were allowed to enter the enclosure to search for caches. On the next two days Orange and Red were allowed to enter individually also to search.

Blue and Green discovered only fifty-two of the 379 caches present when they were allowed to search. About half of these had been made in their presence. Of those fifty-two caches, eleven had been made by Orange, thirty by Red, and eleven by Vander Wall. Overall, Blue and Green did not do very well—their success rates were only 12 and 9 percent respectively. They hopped about, probing the soil with their bills, giving special scrutiny to irregularities of the ground surface.

Orange and Red did far better. Orange recovered twenty-seven caches and relocated forty-four others that had previously been emptied by Blue, Green, or Vander Wall. Significantly, Orange had made all forty-four of those caches.

Red recovered twenty-five caches, and relocated thirty that had been raided by Blue, Green, or Vander Wall. Twenty-seven of those thirty had been made by Red. Orange's success rate was 64 percent, Red's was 78 percent. When Orange and Red foraged, they hopped or flew directly to the caches, made bill probes, removed the seeds, and either ate them or recached them elsewhere.

This experiment showed that only the bird who makes a cache has the necessary information to find it, effectively ruling out the use of scent. Vander Wall also did a statistical analysis which showed that the success rates of Orange and Red were 160 to 190 times greater, and those of Blue and Green twenty to thirty times greater, than would have been achieved by random search, thus ruling out that most improbable of alternatives.

In another experiment that targeted the memory hypothesis, Vander Wall allowed Orange and Red to come in privately and make 151 caches. He then shifted all of the shrubs, logs, rocks, and stumps on half of the enclosure eight inches (twenty centimeters) to the right. The undisturbed half served as an experimental control. During the following week the birds were ushered in to find their caches. Both birds were significantly more successful in finding caches in the control area than in the experimental area. In the experimental area both made numerous digging errors, 69 percent of which were within two inches of a point one foot to the right of the cache site. In other words, Orange and Red had been hoodwinked into thinking their caches were still the same distance from the natural landmarks as they had made them. As Vander Wall laconically put it: "These results are consistent with the predictions of Hypothesis 5." Indeed, he had produced the first scientifically credible evidence that Clark's Nutcrackers use their memory to find

their food caches, and that they do it by using visual cues provided by fixed objects.

Russell Balda and psychologist Alan Kamil showed later that Pinyon Jays and Scrub Jays also relocate their caches by spatial memory, and that the acuteness of that memory increases with the species' dependence on stored food.

In additional experiments Vander Wall explored further complexities and subtleties of nutcracker behavior. The series of experiments failed to produce any evidence that olfaction or random search have a role in relocating caches. It was established that visual cues of microtopography were used by non-caching birds, and to a lesser degree by cachers. But because these cues disappear due to wind, rain, litter-fall, and snow cover, they are useful for only a short time.

Directed random search was used with some success by non-caching birds, who took advantage of the generally greater density of caches near a fixed object than far from one. But caching birds were successful thirty to forty times as often as they would have been under the predictions of this method.

Memory therefore was decisive as the means by which nutcrackers recovered cached food. Vander Wall's experiments conclusively demonstrated that caching birds remember cache locations with respect to objects that serve as landmarks, closer objects being more important than distant ones. Kamil and Balda have produced evidence suggesting that some cache sites are remembered better than are others, and that these are the first ones a nutcracker empties. Their laboratory experiments have also shown that Clark's Nutcrackers become somewhat forgetful of their cache locations after about six months, but that even after nearly ten months they still find their caches more frequently than would be expected by chance.

Perhaps not *all* of their thousands of cache locations will be remembered. The number that saturates a nutcracker's memory—if Sherlock Holmes's "cluttered attic" metaphor holds true here—is unknown. We are only sure that enough caches are found to ensure survival of the cacher and its offspring.

But how can nutcrackers survive when there is a failure of the seed crop? What do they do when the food they are most dependent on is not provided?

CHAPTER 7

Other Arrangements

HEN THE ENLIGHTENED CZAR ALEXANDER II of the Russian Empire ordered "cedar-pines"—*kedr*—to be planted in Finland and Sweden in the 1860s, he had no idea that a century later they would nurture *kedrovka*, the pine nut-eating birds that liven autumn in Siberia.

The 1860s had been years of famine in his western duchies, and the czar's order was intended to provide long-lived trees that would bear nutritious nuts for many years as a hedge against future crop failures. *Kedr* nuts were an esteemed food and a source of cooking oil in Siberia, so it seemed only natural to spread this useful tree beyond its precincts. But in Finland the czar's order was not taken seriously, and resulted mainly in the pro forma establishment of handsome pines for shade in parks, railway stations, and other public places. A seed had been planted, however, and a new tree was introduced into the country.

Eventually, researchers at the Finnish Forest Research Institute's Punkaharju Research Station used seeds of the czar's trees to grow experimental plantings. These thrived at Punkaharju, located in a forest of Scotch pine and silver birch between Lakes Puruvesi and Pihlajavesi, 350 kilometers north of Helsinki and just west of the border of Russian Karelia. Soon the Siberian stone pines began to bear cones. By the late 1930s fourteen forest plantations and numerous rows of roadside trees, a sizeable source of *kedr* nuts, had grown up around Punkaharju.

Nineteen sixty-eight was a hard year for Siberian Nutcrackers. The stone pines were barren that year, and nutcrackers left their breeding range in great numbers, flying without known destination in search of food. A few birds, emaciated and exhausted, even made it to Great Britain, two thousand miles from the Urals, and large flocks were reported in Denmark, Sweden, and Finland. Some of them found Punkaharju.

The discovery by starving Siberian Nutcrackers of Siberian stone pine nuts originally intended for starving Finns was soon noticed by local ornithologists, who welcomed a new species for the life-list. The nutcrackers, who arrived in 1968 under duress, never left. Instead they founded a breeding colony that after two decades contained twenty to twenty-five pairs. These birds have become prominent in the local forest scene, noisily harvesting, transporting, and caching nuts in the fall, unearthing them, and feeding them to their nestlings in the spring.

In September 1989 I visited Punkaharju to assess the impact of these nutcrackers on the composition of the local forest. At numerous locations within a third of a mile from planted Siberian stone pines, research forester Teijo Nikkanen and I found clumped stone pine seedlings that marked the sites of sprouting seed caches. Siberian stone pine was becoming part of the forest flora on the mainland and at least one offshore island. Growing amid the heather, reindeer moss, low-bush blueberry, and sphagnum moss, these pine seedlings will one day take their place in the Scotch pine-silver birch canopy, creating a Siberian forest community a thousand miles from home.

Such eruptions of Siberian Nutcrackers as that of 1968 are by no means rare. According to Claude Crocq, the great naturalist Buffon noted their occurrence in 1754 and 1763, and they have recurred at least thirty times between 1753 and 1977. These mass flights have long stirred the imaginations of Europeans who saw them as portents of war, famine, or epidemic.

Between 1898 and 1964 Clark's Nutcrackers erupted on at least six occasions from the high Sierra Nevada to coastal or desert areas in California. Numerous birds wintered in 1955–1956 on the Monterey Peninsula, where they subsisted on suet provided at feeding stations, insects gleaned from beneath overturned cow dung, yellowjackets, and possibly seeds taken from the massive cones of Monterey pine. It has been suggested that eruptions are most likely to occur when a cone crop failure follows two or more years of cone abundance, because the good years would increase the nutcracker population dangerously, forcing the birds to choose between migration or starvation when the food supply finally crashes. Curt Dimmick has speculated that juvenile birds, which are less efficient cachers and may have inadequate food stored in the soil, may be more prone to erupt than adults.

"Extralimital" migrations like these, in which the birds erupt well outside their normal range, attract the most attention. But nutcrackers probably redistribute themselves within their range far more often in response to more local food shortages.

For example, between mid-August and early October of 1977, many small flocks of Clark's Nutcrackers were seen flying south over the mountains of northern Utah. Most of these birds appeared to winter in pinyon-juniper woodlands in southern Utah, well within the species' normal breeding range. Despite the stress of migrating several hundred miles into a strange area where the terrain, the food source, and the predators presented unknown challenges, some of these birds bred successfully and flew north the following summer with their offspring.

It may have been erupting birds that gave rise to Wilbur C. Knight's observation in September, 1898, of Clark's Nutcrackers "in vast numbers about the towns at the foot of the Wind River mountains (in Wyoming). They remained for the ten days that I was in the locality and fed daily about the back doors of the miners' cabins and became very bold." That boldness may well have been triggered by hunger. I. V. Zykov has recorded a remarkable incident in Siberia that illustrates both the navigational ability and the persistence of hungry Siberian Nutcrackers:

> In the forest-steppes near Tretyakov Lake in a small patch of forest about 10 large cedar-pines had survived. At the start of ripening of cedar-pine nuts at this woodlot kedrovka would appear and with them a struggle set in to save the crop. In 1949 when I visited this place there was a very good yield of cedar-pine nuts. The kedrovka were not tardy in showing up here, promptly at the time of nut ripening. The keeper, a harvester of nuts, engaged the birds in combat. With a rifle he shot over a hundred of them but new birds kept coming in. It was concluded only when the birds had destroyed the crop. I came in on the conclusion of the battle; of cones on the cedar-pines a negligible quantity remained, while through the branches swarmed energetic and voracious kedrovka concluding the "crop-harvest."
>
> It is significant that the nearest forest was about twenty kilometers away from the above-described locality. These many individuals could not have lived in the woodlot. It means they came in from twenty kilometers distance. While they do not transport cedar-pine nuts for such distances it showed their capacity for distant passage and their orientation precisely to a described spot, a small islet of forest vegetation, separated from the forest proper. (Zykov 1953, p. 14)

But if a prize is to be given for the longest precision flight made by nutcrackers to acquire food, perhaps it should be awarded by the state of North Dakota. In 1964 ecologists L. D. Potter and D. L. Green described in *Ecol-*

ogy a 208-acre outlier of limber pine in the Little Missouri Badlands country of Slope County, in the southwestern part of that state. This outlier is over two hundred miles from the nearest major limber pine center, in Wyoming's Big Horn mountains. At that time, pines up to 238 years old were thriving, and the stand was successfully reproducing itself. Potter and Green suspected the limber pine had been seeded "either accidentally or purposely, by Indians camping in the area before settlement," but I suspected Clark's Nutcracker. Diligent efforts by Harry Hutchins turned up an obscure report in *Prairie Naturalist* by L. W. Oring and R. W. Seabloom (1971) of nutcrackers gorging themselves on limber pine nuts in that very stand. The birds stayed for almost a year, and were absent in nearby areas where there was no limber pine.

Food shortages are not always as dire, however, as that in Siberia was in 1968, and nutcrackers do not always need to erupt so drastically into areas outside their breeding range. More often their response to low food availability is to switch to another food species. Clark's Nutcracker can often find a different species of wingless-seeded soft pine. For example, in some years when our study area in Squaw Basin, Wyoming, lacked a whitebark pine cone crop, extensive lower-elevation woodlands of limber pine had cones in abundance. These areas were within five to eight miles, just a few minutes' flying time for a nutcracker, and it is likely that the birds harvesting there were the same ones we knew from Squaw Basin. Limber pine seeds are usually wingless, like those of whitebark pine and are similar nutritionally. Limber pine cones open upon maturing, however, so the seeds must be harvested promptly or they will fall to the ground where nutcrackers seldom forage. Clark's Nutcrackers, who inhabit the Rocky Mountains from the southern extent of whitebark pine in the Wyoming Range north to the Canadian Rockies, can usually choose between these pines, or feed on both. Nutcrackers living in the southern Sierra Nevada mountains can also utilize both limber and whitebark pine nuts, as well as those of singleleaf pinyon, which grows on the eastern slope. And throughout the Great Basin—between the Sierra and the Rockies—they can choose between limber pine and singleleaf pinyon.

When there is no wingless-seeded pine to fall back on, the birds may use a pine with large, winged seeds. Throughout the Sierra Nevada and southern Cascades, the large-seeded Jeffrey pine is available at elevations close to those where whitebark pine occurs. The wide-ranging ponderosa pine grows downslope from whitebark in the Sierra Nevada, Cascades, and northern Rockies. In the Southwest, the preferred species at high elevations is the southwestern white pine, but should its crop fail, the slack can be taken up by seeds from vast woodlands of Colorado pinyon, or middle-elevation stands of ponderosa pine. So Clark's Nutcracker almost always has a choice of more than one wingless-seeded soft pine; or of a large-seeded, winged-seeded, hard pine that grows in the vicinity. Indeed, it has been known to feed upon seeds of all of these pines, and even on the small seeds of Great Basin bristlecone pine.

With the rich pine flora of its western North American range, Clark's Nutcracker seldom has to exploit another genus. There have been no reports of Clark's Nutcracker feeding on spruce, fir, hemlock, or larch seeds, though it does utilize those of Rocky Mountain Douglas-fir in Alberta, Utah, and Wyoming. Perhaps the most vulnerable Clark's Nutcrackers are those of the northern Canadian Rockies and Coast Mountains of British Columbia. Not only do they inhabit an area in which the whitebark pines are few and far between, but they also lack alternative large-seeded species. The only other pines nearby are lodgepole and western white pine, neither of which is known to be eaten by the nutcracker. The other major conifers of those areas are spruces and firs.

Siberian Nutcrackers do not have a wide variety of pines to forage on when the *kedr* crop fails; in the western half of their range there is only the Scotch pine, which has tiny seeds probably not worth the harvest. But unlike their North American counterparts, these birds have learned to forage on spruce seeds, and Russian ornithologists say that major eruptions of this bird occur only in years when both the *kedr* and the spruce are bare of cones. At Punkaharju I watched *kedrovka* take seeds from Norway spruce cones, and a Central Asian subspecies of the Eurasian Nutcracker (subspecies *rothschildi*, the Tian Shan Nutcracker) is said to live entirely on the seeds of Tian Shan spruce, *Picea schrenkiana*, hiding whole cones under the moss (see chapter 12). In the eastern part of their range, Siberian Nutcrackers can choose among several suitable pines. Perhaps the most surprising of nutcracker staples is the hazelnut, which is cached and eaten by populations of European Nutcrackers that live in Swedish forests (where Norway Spruce is also available).

Being able to substitute one species of pine for another may allow a nutcracker to live a relatively undisturbed life when the preferred species is only moderately available. A nutcracker foraging no more than five miles from its communal caching area, after all, can harvest seeds from more than fifty thousand acres, and a ten-mile radius encompasses more than two hundred thousand acres. There might be sufficient cachable food in so large an area to get several pairs of birds through the winter in some lean years, even if that is insufficient for them to raise a brood.

The ability of a nutcracker to migrate when food is unavailable in its home range, or to switch foods when its preferred food is in short supply, shows that it is not completely dependent on any single species of pine. But what of the dependency of a pine on the nutcracker? From the standpoint of white-bark pine, for example, is Clark's Nutcracker merely a convenient means of seed dispersal, or is it a necessary agent of both seed dispersal and the establishment of new seedlings?

We sought the answer to this question in the meadows and forests of Squaw Basin.

CHAPTER 8

Who Needs Clark's Nutcracker?

Y THE LATE 1970s IT WAS CLEAR THAT SEVERAL pines in the Old and New Worlds owed at least some of their regeneration to corvids. A long tradition of such observations could be found in ornithological reports from the Alps and Siberia that acknowledged the Eurasian Nutcracker's role as an establisher of stone pine seedlings by virtue of germination in seed caches. European investigators, for example, had described the European Nutcracker as "an essential help in reforestation," without which stone pine regeneration is "scarcely conceivable." Similar findings were beginning to appear from western North America. An important paper by Russell Balda and Gary Bateman of Northern Arizona University provided evidence of large-scale caching of Colorado pinyon seeds by Pinyon Jays. Studies by J. David Ligon at the University of New Mexico provided physiological information about the jay that helped explain its behavior as a pine-nut bird.

Stephen Vander Wall, working under Balda's supervision, examined the relationship of Clark's Nutcracker and Colorado pinyon, and made observations on the utilization of pine nuts by Scrub and Steller's Jays as well. An innovation of Vander Wall's was an attempt to explain the morphology of pinyon cones and seeds as evolved characteristics resulting from the mutualistic interaction. In the Sierra Nevada, Diana Tomback studied the foraging of Clark's Nutcracker in high-elevation forests of whitebark pine. Concurrently, Hermann Mattes was making studies of the European Nutcracker in the Swiss Alps, elaborating on earlier work by Friedrich-Karl Holtmeier, and Claude Crocq was doing the same in the Maritime Alps of southeast France.

As often happens in science, the impetus for much of this work seems to have been a trail-blazing review article, "Ecological Aspects of Food Transportation and Storage in the Corvidae," by Frantisek Turček and Leon Kelso, which appeared in the journal *Communications in Behavioral Biology* in 1968. Turček and Kelso reviewed the reports of more than 175 studies relating to corvid food storage behavior. Written in English, Polish, French, Bulgarian, Finnish, German, Slovak, and Russian, and dating back as far as a hundred years, many of these reports appeared in obscure journals. Turček and Kelso's review showed a dramatic spurt in pine-corvid studies in the Soviet Union during the 1950s and 1960s. They assembled data from old and new studies and drew some broad conclusions on the importance of food storage in the ecology and evolution of the Corvidae. Turček and Kelso, like most of the authors they cited and most of the researchers named above, were bird biologists. Their major focus was the behavior of corvids vis-à-vis the pine nut food source—harvest, storage, and retrieval—and its ecological and evolutionary repercussions on the corvids. Foresters had hardly become aware of these mutualisms, and some Russian foresters considered *kedrovka* a seed predator without redeeming character.

I first learned of Balda's and Bateman's (1971) work on the relationship of jays and pinyon pines when I started to write a book on those pines late in 1973. The following fall Stephen Vander Wall, who had just submitted his study of Clark's Nutcrackers and Colorado pinyon for the master of science degree at Northern Arizona University, appeared in my office at Utah State University to solicit information on local pinyon woodlands. Vander Wall had just moved to Utah to begin work on the Ph.D. Upon reading his thesis draft, and especially its references to European and Siberian research, it struck me that corvid-pine mutualisms must partially answer a rhetorical question posed a decade earlier by my friend, the late William Critchfield, a leading authority on pine genetics and evolution. Critchfield had asked, "Isn't it strange that so many white pines live in stressful habitats—semi-arid or alpine?" In the sort of intuitive flash that I have experienced only half-a-dozen times in my research career, I knew (with a certainty that usually lasts only until the facts come in), that *wingless* seeds indicate corvid dispersal and establishment, that the large size of those seeds is adaptive to harsh environments, and that corvids had exerted the selection pressure that lay behind the evolution of those seeds. Clearly, this was worth some research! I tentatively decided to concentrate my efforts on the poorly understood whitebark pine, and in a field trip to the Rockies of British Columbia in 1975, I made some preliminary observations. Other research obligations interfered with whitebark pine plans until early in 1977. By then I had realized that my own interests in whitebark pine could not be realistically pursued until the basic "bird work" was done. A competent ornithologist would first have to establish that Clark's Nutcracker did indeed harvest, cache, recover, and feed upon whitebark pine seeds as Vander Wall had demonstrated for pinyon. I had neither the training nor the skills for that. David and Martha Balph, colleagues of mine at Utah State, and recognized bird behaviorists, agreed to collaborate with me by doing the much-needed bird work. Shortly afterward, the Balphs attended a meeting of the Cooper Ornithological Society. On their return, they described how a young Ph.D. candidate from California, Diana Tomback, had reported on foraging studies of Clark's Nutcracker on whitebark pine in the Sierra Nevada. "She's done everything we were going to do, so there's no need for us to do it," Dave told me. My work could now proceed more rapidly. Tomback's research, which was published the next year (1978), showed that nutcracker utilization of whitebark pine seeds was very similar in virtually all respects to what had been reported in fragmentary form with regard to Swiss and Siberian stone pines by many European authors, and in detail with regard to pinyon by Vander Wall. This gave me the needed proof that whitebark pine was indeed a bird pine, like its close relatives in the Alps and Siberia. A coherent analysis could now be made of pines with wingless seeds. In the summer of 1978 I received a grant from the National Science Foundation to pursue these relationships from the focal point of the pine species.

As a forester, I was especially interested in learning the relative importance of Clark's Nutcracker to whitebark pine. Is the relationship a luxury, or a necessity? Is Clark's Nutcracker the unique dispersal and establishment agent for this lovely pine, or is it merely one of a number of animals that does the right thing?

Strict one-to-one dependence of a plant on an animal and vice versa—an obligatory mutualism—is very rare. Far more common is reliance on a variety of species, by both the plant whose seeds are being dispersed, and by the vertebrates doing the dispersing. For example, in the tropical dry forest of Santa Rosa, Costa Rica, the oily, red arils of the small tree *Caesaria corymbosa* attract at least fourteen bird species. Four of these—the Yellow-green Vireo, Streaked Flycatcher, Golden-fronted Woodpecker, and Pale Flycatcher—remove about nine-tenths of the seeds and disperse them throughout the area. At least eight other birds harvest the remainder of those seeds. In the pinyon-juniper woodland around Flagstaff, Arizona, Michael Salomonson identified six wintering bird species and four mammals as probable dispersers of one-seed juniper; and as we saw earlier, Colorado pinyon has four corvid dispersers.

Many of our temperate zone fruits mature in time to attract fall migrants, thus making themselves available to large numbers of birds of diverse species. Pedro Jordano found in southern Spain that fall-fruiting blackberries were fed upon by twenty species of migrant birds who dispersed the hard, little seeds in their droppings. These included Blackcaps, Whitethroats, Sardinian and Garden Warblers, and even Pied Flycatchers. In all these cases, not only is the plant not totally dependent on any one bird, but none of the birds are totally dependent on blackberries.

The same generality is apparently true of bats. The Short-Tailed Fruit Bat (*Carollia perspicillata*) of Central and South America has been shown to feed on the fruits of eight shrub and seven tree species in the tropical dry forest of Costa Rica. Nearly all of those plants' fruits are also eaten (and the seeds consequently dispersed) by two to nine other bat species, as well as other mammals and birds. Fruits of three of the tree species are consumed by more than twenty animals.

On the other hand the Tambalacoque tree, *Calvaria major*, which grows only on the little Indian Ocean island of Mauritius, apparently coevolved with the Dodo bird. The Dodo fed on *Calvaria* fruits, and passed the hard-coated seeds, which were prepared for germination by the mechanical abrasion they underwent in the Dodo gut. No other bird provided Tambalacoque this service. Around 1675, the last Dodo died. Tambalacoque trees ceased to be replaced by new seedlings, and by the late 1970s only a few old trees remained. The Dodo was taking Tambalacoque with it to extinction. Is the whitebark pine a potential Tambalacoque?

Further, if whitebark pine could only be dispersed and established by Clark's Nutcracker, then its geographic range, site occurrence, spacing, successional status, and the genetic architecture of its populations would all be hostage to this bird's behavioral quirks. And the morphological characteristics of the pine, its cone structure for example, as well as its physiological tolerances, could be products of the selective pressures exerted by the nutcracker in its harvesting and caching activities. But if several animals dispersed and established the pine, its characteristics could reflect the behaviors of them all. Comparison of distribution maps showed that everywhere that whitebark pine grows, Clark's Nutcracker is also present, but this was more suggestive than conclusive.

Several researchers had implied that nutcrackers were mainly or exclusively responsible for seed dispersal in stone pines. But none had produced data that excluded other animals as possible dispersers. In short, it had become obvious that nutcrackers were whitebark pine nut dispersers, but it was not obvious that they alone performed this service.

A confusing factor that had to be dealt with was the "rotting cone hypothesis." Some earlier investigators of whitebark pine had claimed that the tree shed its cones in the fall, while they were still closed and full of seeds; that the cones then lay on the ground until they decayed; and that at some unspecified time afterward, the seeds fell out of the rotten cones and germinated. This idea, though encountered repeatedly, was unsupported by any data. But it did need to be addressed because, supported or not, it had precedence over any new explanation. It is far easier to establish misinformation in the scientific literature than to dispel it after it has become entrenched.

We doubted that mature cones were shed, because during several years of field work we never saw any fall from the treetops other than through animal action.

In the summer of 1978 I covered forty-six maturing cones with bags of heavy nylon mesh to protect them from nutcrackers and squirrels. By the following July, twenty-five of these cones had been destroyed or removed, their bags riddled and in shreds, nine lay loose in the bags, and twelve were still attached to their branches. The latter group yielded 347 seeds, of which 321 (93 percent) were rancid (brown and shrunken) and inviable, presumably from basking in the hot summer sun. By contrast, of 112 seeds stored under an inch of soil during the same period, only 9 percent were rancid. Suspecting that the cones that were loose in their bags had been detached by animal action or handling, I attempted another trial the next year, using a double-thickness of nylon over fifty cones. But by October, nutcrackers and squirrels had already destroyed the bags, and shredded or taken away every cone. More studies of this kind are needed. While I remain unconvinced that mature, seed-filled cones are normally shed, I will admit to one exception. In the fall of 1989 there was a massive crop of whitebark pine cones throughout the subalpine forests of western Wyoming and Montana. In the groves on the Squaw Basin meadows, the cone crop was so far out of balance with the nutcracker and squirrel populations that in July 1990 there were still thousands of intact, seed-filled cones in the trees. Though nutcrackers dug up cached seeds, they ignored the cones persisting in the tree crowns. These cones needed only the lightest of touches to dislodge them. They were light brown in color with dry white incrustations of resin on their scales. Analysis of ten cones picked from trees and ten cones recently fallen to the ground revealed that 69 percent and 57 percent of the seeds respectively had rancid kernels. Hermann Mattes has observed that the cones of Swiss stone pine are also securely attached, and fall from the tree only in the spring following a cone crop too large for nutcracker populations to cope with.

Nor is it likely that large, nutritious pine nuts can lie on the soil surface long enough to germinate without being devoured by a hungry rodent. During the 1978 limber pine mast year, reportedly the best crop year in decades, I set out wire-mesh trays of forest-floor litter seeded with limber pine nuts beneath fruiting limber pines in Logan Canyon, Utah. Of one thousand seeds (from forty trays in two locations) only three were still intact after just nine days. The trays were littered with seed-coat fragments, many of them with plainly visible tooth marks. A later trial with whitebark pine nuts in Squaw Basin yielded similar results. None of sixty placed on the soil surface in September were still present the next July. In a second test only 7 percent were intact after only two weeks. By contrast, 10 to 43 percent of seeds sown at nutcracker cache depth (three centimeters, or just over one inch) were still present the next June when hidden in the forest; and 63 to 100 percent survived in caches made in the meadow. Similar results have been reported by Ward McCaughey, who lost 100 percent of broadcast whitebark pine seeds to rodents. These results show that seeds lying on the surface quickly become rodent food. Those that are cached in the forest suffer heavier predation (probably from squirrels) than those in the open, but many of them are likely to survive and germinate. In other words, surface seeds perish, while buried seeds can survive. So much for the rotting cone hypothesis.

Wind dispersal could be dismissed out of hand. A few years earlier a student of mine, Ronald Warnick, had done calculations showing that a singleleaf pinyon seed falling from a thirty-foot height would be carried only thirty feet from the tree by a hundred-mile-per-hour wind. Moving wingless pine seeds by wind is obviously no way to run a forest.

Therefore, animal activity was surely the agent of whitebark pine seed dispersal. By which animals? The only way to definitively answer this question was to spend many, many hours in the field while a cone and seed crop was maturing, and to study in detail the behavior of every animal that foraged in the treetops. That was Harry Hutchins's assignment.

Hutchins's strategy was to make weekly rounds of several thousand acres of forest and meadows in Squaw Basin, a 9,300 to 9,600-foot-high area in Wyoming's Bridger-Teton National Forest. The basin's rolling meadows contained numerous low-lying swales, wet after snowmelt, which supported typical herbaceous plants. Brushy, moose-browsed willows lined the banks of Blackrock Creek. Gently sloping moraines rising between the swales supported a growth of sagebrush and grass, and some of these were crowned with scattered to dense-grown whitebark pines of various sizes and ages. On others, and on ridges surrounding the basin, there were thick stands of Engelmann spruce, whitebark pine, and subalpine fir. Harry cruised the dense forests and the open woodland-like areas routinely to track the maturation of cones and seeds on selected trees, and to determine the rate at which seeds were being depleted by animals throughout the summer and fall. As he passed through the area he would stop to identify all diurnal vertebrates in the crowns of whitebark pines, and, with a stopwatch, measure the time they spent in seed harvesting, seed caching, cone harvesting, cone caching, feeding, seed dropping, flight, preening, play, aggression, and resting.

At first Harry's task was almost idyllic: he freely roamed the scented forest and flowered alpine meadows, making notes on the birds and small mammals that fell into his sensory net, tanning in the sun beneath an azure sky. The bubble burst early in August when he came upon a great mound of fecal matter that told him he was sharing his research area with a very large quadruped. Bear experts Barrie Gilbert and Fred Lindzey back at Utah State University confirmed the correctness of the hairs-on-the-back-of-the-neck diagnosis Harry had made in the field—an interested Grizzly Bear was doing research of its own. Thereafter Harry traveled less frequently on foot and more often in a four-wheel-drive vehicle, and he persuaded his fiancée, Sue Nelson, to accompany him on his rounds, alternately shouting and shaking a cowbell to give all grizzlies within earshot fair warning of their presence.

Many of the vertebrates Harry observed in close proximity to whitebark pine trees made no attempt to harvest seeds. These included the Common Flicker, Cassin's Finch, Rosy Finch, Pine Siskin, Dark-eyed Junco, the Pine Marten, and Weasel. To our mild surprise the Gray Jay was never seen foraging on whitebark pine seed during 3,500 seconds of foraging observations. Gray Jays did indeed cache food, but only in the form of fresh carrion, or boli stuck with saliva to conifer branches. Two stray Magpies flew through the basin without stopping. Coyote scats showed no evidence of pine seeds. These animals, then, were mere bystanders in the whitebark pine's life history. But a surprising number and array of vertebrates did make the list of potential dispersers.

CLARK'S NUTCRACKER

Clark's Nutcracker, to nobody's surprise, was by far the most frequently observed resident vertebrate that foraged in the whitebark pines. Nutcrackers foraged far more than any other animals, and their foraging was most often successful; that is, they actually acquired the seeds they reached for (Table 8.1). Almost all of their foraging was in the trees. While pine nuts were available, nutcrackers spent only 1 percent of their foraging time going after insects; and they never foraged on seeds of the other conifers, lodgepole pine, alpine fir, or Engelmann spruce. Clearly, they preferred pine nuts. Nutcracker foraging began at Squaw Basin on August 4, when the mushy, imma-

Table 8.1 Time spent by birds and mammals foraging on whitebark pine seeds as a percent of their total foraging time in whitebark pine stands in Squaw Basin, Wyoming, August 15 to October 11, 1980.

Animal species	Total foraging time (seconds)	Percent of time spent successfully foraging on whitebark pine seeds		Percent of time spent unsuccessfully foraging on whitebark pine seeds[a]
		on trees	on ground	
Clark's Nutcracker	42,401	97.5	0.2	1.3
Steller's Jay	2,541	24.5	14.0	6.9
Raven	572	78.7	0.0	21.3
Pine Grosbeak	1,797	91.7	0.0	2.3
Mountain Chickadee	1,235	7.6	0.0	13.0
Red-breasted Nuthatch	262	0.0	0.0	100.0
Red Squirrel	852	60.0	15.8	0.0
Chipmunk	1,625	35.4	20.8	8.1

Source: Adapted from Hutchins and Lanner (1982).
[a]Unsuccessful foraging results in dropped seeds.

ture seeds could not yet be removed whole from the cones, and ended in early November, when the last of the seeds were eaten or cached (Figure 8.1). By winter, each nutcracker had harvested about 129,000 seeds (Table 8.2), of which 76 percent were stored in the soil. Caches were made at the bases of trees or smaller plants or rocks, in dense moss cover, or in the open among sagebrush and grass. They were made on all exposures—north, south, east, west—near a spring, on a streambank, even in a puddle.

Groups of ten to fifteen nutcrackers cached at the same time within an area of about one hundred square meters, and showed no signs of aggression toward each other. If we had seen no other data on Clark's Nutcracker, these observations would have convinced us of its effectiveness as a disperser and establisher of whitebark pine. The disperser role was served by the re-caching of locally cached seeds to communal caching areas over two miles away, on the Breccia Cliffs. The establisher role stems from the habit of caching seeds in the soil at about the same depth a nurseryman would sow them. As we have seen, soil caches are safe havens for whitebark pine seeds.

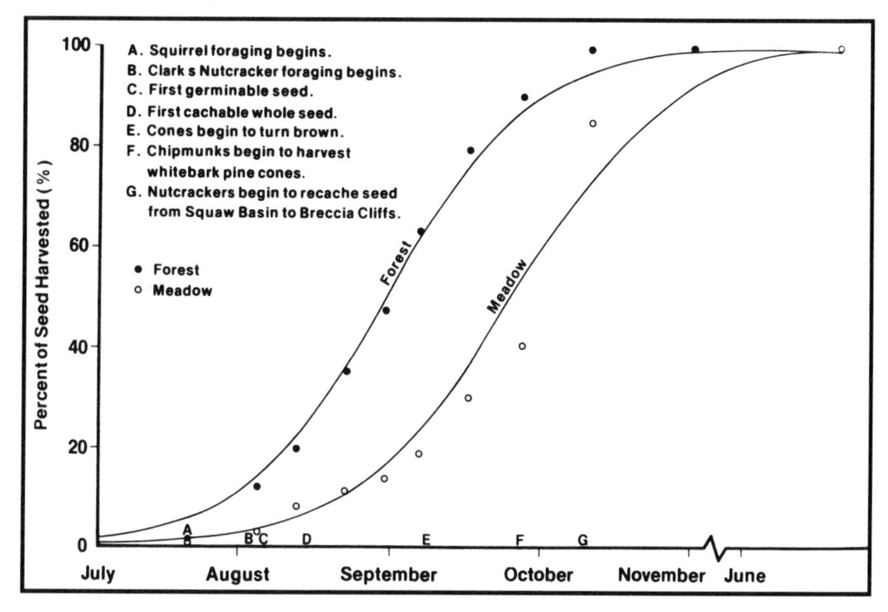

FIG. 8.1 Fate of the 1980 whitebark pine nut crop in Squaw Basin, Wyoming. Trees growing in dense forest were subject to heavy seed predation by Red Squirrels, thus were depleted more rapidly than meadow trees.

Table 8.2 Comparative harvest of whitebark pine seeds by vertebrates in Squaw Basin, Wyoming, August 15 to October 11, 1980, calculated from time-budget data.

Species	Time spent foraging (min/day)	Harvest duration (days)	Seeds harvested per individual	Relative abundance (%)	Relative number of seeds harvested (seeds/ 1,000,000)[a]
Clark's Nutcracker	180	91	129,402	70	364,300
Steller's Jay	120	55	4,620	2	370
Raven	30	53	954	2	76
Chickadees and Nuthatches	120	56	4,700	7	1,320
Red Squirrel	240	84	875,000	18	633,041
Chipmunk	120	35	7,140	2	575

Source: Adapted from Hutchins and Lanner (1982).
[a]Calculated by multiplying the preceding two columns and adjusting to 1,000,000 seed base.

STELLER'S JAY

These elusive birds—dark plumaged, often silent, usually alone or paired—were far less common in the basin than nutcrackers. They arrived late, in early September, and devoted only one-fourth of their foraging time to whitebark pine seeds. Because they were often unable to remove cone scales with their small bills, they tended to forage on seeds previously exposed by nutcrackers. When they did acquire pine nuts, they usually tore apart the contents in eating them, or flew off with no more than five seeds in the esophagus. They were seen caching pine nuts under lichen growth on branches, or in a branch crotch, but never in the soil. Thus while they do have a limited capacity to disperse seeds, their caching habits do not create the necessary conditions for whitebark pine seedlings to become established. Since all the seeds it acquires will die, Steller's Jay must be considered purely a seed predator.

RAVEN

Improbable though it seems, Ravens occasionally forage on whitebark pine seeds. These big-bodied, massive-billed corvids perched above the cones, reached down to pull off some scales, finally acquired one or two seeds, and glided off to a rocky south-facing slope. They were seen caching carrion, but not seeds, under small rocks. The infrequency of their visits, and the lack of caching evidence of the rare seeds they acquire, justify considering the Raven irrelevant to the life history of whitebark pine.

OTHER PASSERINES

Pine Grosbeaks arrived in September in small, wandering flocks. They acquired whitebark pine seeds by removing cone scales, or took some previously exposed by nutcrackers. They tore apart the seeds they ate, and were not seen caching. Mountain Chickadees foraged mainly for insects, but during the pine-nut season they spent some of their foraging time on whitebark pine seeds. They were unable to handle the large seeds deftly, and often dropped them. They tore apart the ones they ate, and were not observed making caches. Red-breasted Nuthatches occasionally foraged among whitebark pine cones, but were not seen successfully harvesting any seeds. All of these birds fail to qualify as potential dispersers or establishers of whitebark pine.

Plate I.1 ABOVE Whitebark pine in Squaw Basin, Wyoming *(David Lanner).*

Plate I.2 RIGHT Clark's Nutcracker testing a limber pine seed for soundness by clicking it in the bill *(R. Lanner).*

Plate II.1 ABOVE Clumped whitebark pines in Squaw Basin, Wyoming *(David Lanner).*

Plate II.2 ABOVE, RIGHT Immature whitebark pine cones riddled by Clark's Nutcrackers *(Utah State University Photography Services).*

Plate II.3 LEFT Whitebark pine is a pioneering species on high-elevation burns in the northern Rockies, Shoshone National Forest, Wyoming *(R. Lanner).*

Plate II.4 ABOVE Grizzly Bears in the greater Yellowstone ecosystem raid squirrel middens for fat-rich whitebark pine nuts *(Barrie Gilbert).*

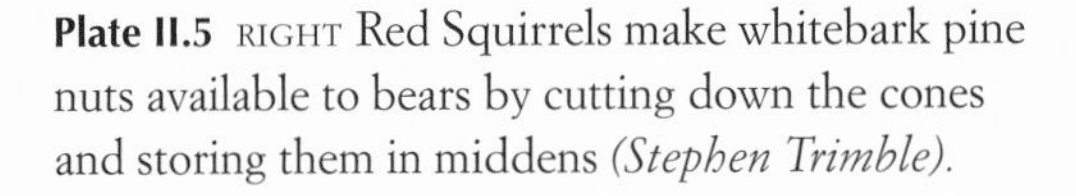

Plate II.5 RIGHT Red Squirrels make whitebark pine nuts available to bears by cutting down the cones and storing them in middens *(Stephen Trimble).*

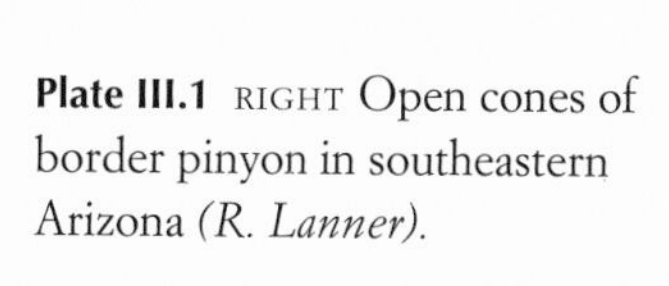

Plate III.1 RIGHT Open cones of border pinyon in southeastern Arizona *(R. Lanner).*

Plate III.2 BELOW The Pinyon Jay is the major disperser of pinyon pine seeds *(Stephen Trimble).*

Plate III.3 ABOVE Eurasian Nutcracker harvesting seeds of Swiss stone pine *(H.C. Richter).*

Plate III.4 LEFT Colorado pinyon in southern Utah *(R. Lanner).*

Plates IV.1, IV.2 Young (above) and ancient (left) Great Basin bristlecone pines are often found in clumps growing from nutcracker caches. White Mountains, California and Spring Mountains, Nevada *(R. Lanner)*.

Plates IV.3, IV.4 Conventional white pines like eastern white pine (above) and western white pine (left) have horizontal branches and dangling cones. Adirondack Mountains, New York and Olympic Mountains, Washington *(R. Lanner)*.

CHIPMUNKS

These little rodents spent most of their time close to sagebrush, and visited whitebark pines infrequently. Observation of these animals in 1980 and 1981 convinced Hutchins that they prefer to feed on such plants as lupines and fireweed, until these die back in the fall. Just over one-third of chipmunk foraging time was devoted to whitebark pine trees, most of which resulted in harvesting success. Chipmunks were spotted climbing the trees starting in late September (Figure 8.1), when the cone scales were loosening. They were not seen caching seeds, and inspection of three burrows to about eight inches from the entrances revealed no stored seeds. Chipmunks are potential dispersers of seeds over very short distances. But since their burrows average almost a foot in depth it is most unlikely a seed germinated there could become a sunlit seedling. Though several earlier chipmunk investigators had asserted that chipmunks are important agents of afforestation, we found no evidence of that being true of whitebark pine.

GOLDEN-MANTLED GROUND SQUIRREL

This attractive little mammal is absent from Squaw Basin, but several were seen foraging on cones in whitebark pines on Mount Washburn, in Yellowstone National Park. They cheek-pouched some seeds and ate them upon reaching the ground. No caching was observed, and this animal, which seldom climbs trees, is considered an occasional whitebark pine seed predator.

RED SQUIRREL

Red Squirrels are very common in subalpine conifer forests, where they make their presence known with loud chattering and other vocalizations. In 1978, when there was a major whitebark pine cone crop in the Rockies, Kate Kendall noted that the Red Squirrels ignored cones of other species and perhaps even mushrooms. When the 1979 cones were available, the squirrels harvested them, even though lots of 1978 cones still reposed in their middens. Harry Hutchins witnessed another manifestation of the squirrels' interest in whitebark pine seeds, in mid-July 1980:

> [Red Squirrels] would frequently retrace a nutcracker cache area after the bird had left. These rodents would sniff the ground diligently in

> hopes of locating a cache of seeds that a nutcracker had missed. I timed one squirrel that sniffed the ground for almost 7 minutes before it gave up trying to find a nutcracker's cache. (Hutchins, unpublished report)

It was therefore no surprise when the data showed them second only to nutcrackers in their whitebark pine visitations. Red Squirrels spent 60 percent of their foraging time on whitebark pine cones and about 11 percent on Engelmann spruce cones (Table 8.1). About one-sixth of their foraging time was spent gathering up from the ground whitebark pine seeds that had been dropped by other animals. Their foraging rates were much higher than those of other animals because they usually harvested an entire cone full of seeds as a unit, especially in August and early September, when the cones could be handled without being broken apart. They started their cone harvest in Squaw Basin on July 22, long before the nutcrackers (Figure 8.1). As the season progressed they added to their harvest cones of Engelmann spruce (mid-August), alpine fir (mid-September), and lodgepole pine (late September).

Squirrels hoard their cones on top of, or deep within, "middens"—areas of the forest floor that are often piled deeply with the cone debris of many years, perhaps several decades. But after mid-September, squirrels began also caching *seeds* from some of these cones. Typically, a squirrel would remove a seed from a cone, run with it to the midden, place it in a deep hole, and then go back to the cone for another seed. This slow and tedious process was most inefficient, taking more than a minute for each seed cached. The number of seeds per cache ranged from fourteen to fifty-five in the four Hutchins observed being made, although Kate Kendall, in her studies of Red Squirrels in Yellowstone National Park two years earlier, counted as many as 176 seeds in a cache.

The squirrel proved to be the most assiduous harvester of whitebark pine seeds within the dense forest, accounting for 875,000 seeds per animal, almost seven times as many as a nutcracker harvested. Out in the meadows however, where the squirrels never trod, the nutcrackers had virtually all the whitebark pine crop to themselves. We surmised that the ever-present Red-tailed Hawks soaring overhead were in large part responsible for the squirrels' discretion. Clearly, we had to give careful consideration to the Red Squirrel as a potential disperser of whitebark pine seeds, and an establisher of its seedlings.

Surprisingly, despite its incredibly high level of activity in harvesting and caching cones and seeds, the Red Squirrel could not be assigned any significance as disperser or establisher. Dispersal requires transport over some distance, but squirrels, despite their frenetic levels of activity, do not go very far.

According to squirrel biologist Christopher Smith (1970), their territories, which they aggressively defend from interlopers, and where they spend nearly their entire lives, range from about one-half acre to three acres in size. Given their limited cruising range, Red Squirrels cannot disperse seeds very far from the mother tree.

The squirrels were equally ineffective as whitebark pine seedling establishers. Nearly all of their seed handling occurred on their middens, yet a careful survey of twenty-five middens, with a total surface area of about 0.4 acres, showed a marked deficiency of young whitebark pine growing on that surface. Assuming the middens to be fifty years of age, then the whitebark pines less than nine centimeters in diameter growing on them can be assumed to have taken root on the midden. Trees larger than that would have been present before the midden was started. There were only twenty such whitebark pines on 250 square meters of midden surface, but 191 on an equal area of randomly located forest plots. In other words, even if all those trees now on the middens have indeed become established on midden surface, this surface was only about one-tenth as productive in producing whitebarks as was the normal forest floor. Actually, some of these trees were probably established on the forest floor, and were later surrounded by a growing midden. The mass of loose debris in a midden—cone scales and cores in varying stages of decomposition—is simply not a good seedbed. It dries out in the summer, and is constantly dug into and churned up whenever a squirrel caches or retrieves a cone. Even the seeds that are buried after being removed from the cone are unlikely to germinate successfully, because they are buried too deeply to emerge. Any that did sprout would be likely prey for squirrels, which were seen on several occasions eating germinating seedlings. We therefore concluded that while the Red Squirrel might occasionally cause a whitebark pine to become established, this would be a random and infrequent event that could have only a minute impact on the pine's population biology.

GRIZZLY AND BLACK BEARS

Both bear species raid squirrel middens to forage on whitebark pine seeds. Limited observations of fecal material of a Grizzly in Squaw Basin disclosed that only one of several thousand seeds was still intact. Four fecal samples on Mount Washburn contained two intact seeds. A mass of Black Bear feces in Squaw Basin contained one undamaged seed. Our limited experience with seed-eating bears suggests these beasts play no significant role in dispersing and establishing the pine. For one thing, very few seeds survive the chewing action of their molars. Those that do survive seem unlikely to put down roots through a fecal pile into the soil below. And since the scats are usually close to the raided squirrel midden—within twenty-five meters in our experience—any such germinants would probably find themselves in the shade of the forest canopy. Further, as Kate Kendall has explained, Grizzly Bear scats contain large amounts of undigested food, including nuts and berries, which attract birds and small mammals, much as grass seeds in horse manure attract sparrows. Kendall thinks an intact seed would probably be eaten by one of these scavengers long before it could germinate.

The results of Hutchins's intensive observations clearly show that Clark's Nutcracker is an effective disperser and establisher of whitebark pine, and that none of the other animals are. The dependence of whitebark pine on this corvid is almost complete. Without the Nutcracker's services the pine would survive only through improbable establishment events—an unusually shallow squirrel cache allowing germination of an occasional seed here, a seed overlooked by foraging rodents managing to germinate there—but systematic, reliable, or large-scale establishment would be impossible. A species cannot survive through the kindness of fortuity. So for whitebark pine, Clark's Nutcracker is no mere convenience, and surely not a luxury—it is the indispensable giver of life. But perhaps even more remarkable is the nutcracker's role in shaping the ecosystem of which whitebark pine is only one member.

CHAPTER 9

Building Ecosystems

HE TRIGGERING OF GREAT EVENTS BY SLIGHT force is a common theme in science. Thus very small amounts of a catalyst can allow a chemical reaction to go forth which would otherwise not occur. The amount of reactive matter greatly exceeds the bulk of the catalyst, which is itself unchanged by the reaction. In the realm of biochemistry an enzyme, which is a protein of very specific function and molecular structure, acts much like the catalyst in an inorganic chemical system, controlling the rate at which chemical reactions progress in a much greater mass of substrate. And at the level of the whole organism, hormones formed in the leaves or roots of a plant, and present in very low concentrations, are transported to sites where they stimulate growth or movement of relatively large masses of plant material. The amount of the hormone indoleacetic acid (IAA) in the first leaf (coleoptile) of an oat seedling, for example, is so slight that an estimated twenty thousand tons of coleoptiles would be needed to extract one gram of the hormone. Yet even at this remarkably low concentration, the IAA controls coleoptile growth. In all of these examples there is a striking disparity between the amount of an active substance and the ultimate impact of that substance on its milieu. At an ecosystem level the same can be said of the effects of a seed-caching, seedling-establishing animal like the nutcracker.

In and around our Squaw Basin study area there are about five thousand acres, most of it forest, inhabited by about one hundred Clark's Nutcrackers. The actual home range of this flock is probably far in excess of five thousand acres, but virtually nothing is known of this aspect of nutcracker natural history. The biomass (dry weight of organic material) of the whitebark pines on that conservatively-drawn area probably amounts to several thousand tons. The biomass of the rest of the forest may be in the tens or hundreds of thousands of tons. Yet a flock of nutcrackers, boasting perhaps thirty-five pounds of fresh weight, catalyzes changes in the vegetation and animal life of those thousands of acres, and determines where much of the biomass will develop for centuries to come.

Plant ecologists refer to natural changes in the makeup of plant communities as plant succession. The study of plant succession is almost as dynamic as are the changes that are studied, and a large underpinning of theory has been invented to explain the complexities inherent in the process. But let us oversimplify.

The establishment of plant cover on a formerly unvegetated surface is termed primary succession. The changes in species composition that occur on an already-vegetated area comprise secondary succession. The plants that participate in primary and secondary successional processes are often of different species, and in any successional process we expect different species to be found in early and late stages, or seres. Succession can be set back when a disturbance of some kind eliminates the plant cover, setting the stage for earlier-successional species to come back in. In the absence of disturbance, a relatively stable forest will develop, perpetuating itself until a catastrophe comes along. Within a forest, the ability to tolerate shade is one of the crucial determinants of a species' successional status, with early pioneers exhibiting low shade tolerance, and late seral species showing considerable tolerance.

An example: in the North Carolina piedmont, loblolly pines invade open fields abandoned by agriculture. They grow up in an almost-pure stand, and all the trees are about the same age. They thrive as the dominant trees of the sunlit overstory. Beneath them, however, few loblolly seedlings survive, because they cannot tolerate the shade of the overstory or the competition for water in a soil already filled with pine roots. But oaks can tolerate these conditions, and they develop beneath the pines, awaiting the time when the pines pass on, leaving the oaks as the dominants of a new overstory.

Pines are almost always early-successional, or early-seral. They often occupy abandoned fields, cutover forests, burned areas, raw soils of glacial outwash plains, and even beach sands. They are not expected to grow in the shade. That is for the oaks, maples, spruces, or firs that will eventually come up from below to replace them. The pines usually form the first wave of forest succession: they grow up, they grow old, they die, and they make way for a second wave of a more shade-tolerant species.

But it does not necessarily work that way when nutcrackers make the decisions about where seeds will be placed, and when. In Squaw Basin whitebark pine is indeed an early-successional species. It not only pioneers on the moraines that rise above the meadows, but on the bare, rocky ledges of the Breccia Cliffs as well. Yet it is also late-successional. Forests ringing the basin consist mainly of Engelmann spruce (*Picea engelmannii*) and subalpine fir (*Abies lasiocarpa*), with old whitebark pines that were there when the spruces first arrived. In addition, there are high densities of whitebark pine seedlings and saplings slowly growing in the shade of the dense overstory, each waiting patiently for the opportunity to grow up into the canopy gap left by a dead or fallen tree.

The forest is moving out into the meadows of Squaw Basin by first colonizing the tops of isolated moraines, and then expanding on the high ground until the new patches of forest coalesce. A study by Karen Snethen (1980) has filled in some of the details of that process.

Snethen had observed that many of the rocky moraines rising from the meadows of Squaw Basin were crowned with groves of whitebark pine. Some groves consisted of only a few small saplings or seedlings, while others had many trees ranging in size from seedlings to trunks several feet thick. It appeared that the older the grove, the more likely it was to have an admixture of Engelmann spruce.

Snethen selected eleven discrete groves, ranging in size from ten to more than a thousand trees, and measured their diameters as a rough index of tree age. These groves formed a "chronosequence"—a series of tree populations ranging widely in age and assumed to reflect the changes that would occur in a single stand over its lifetime. The youngest population consisted of just six seedlings and saplings of whitebark pine, the oldest of which was about sixty years old. There were several other pure stands of whitebark pine, the oldest with trees more than 120 years old. Beyond that point, spruces began to appear among the whitebark pines, at first represented mostly by seedlings and small saplings, but eventually by larger trees as well. In the oldest of these miniature forests Karen found that both pines and spruces were present as seedlings, saplings, pole-sized trees, and mature trees up to thirty-two inches in diameter. In a typical old stand the whitebarks were two to three times as old as Engelmann spruces of the same diameter. The oldest whitebark pine, which was over 295 years of age, was already about 144 years old when the first surviving spruce seeded in.

The open meadow is a harsh environment in the summer, when the hot midday sun bakes the ground, and the breeze sucks moisture from plants and soil. The tiny, winged spruce seeds that rain on the meadow in the fall only rarely germinate and survive long enough under these conditions to become established seedlings. Whitebark pine has a far better chance to reach treehood. Its seeds are buried by nutcrackers, greatly reducing the threat of dehydration and rodent predation. The seeds are large, so the new germinant has abundant food stores to draw upon, especially in building a young root that can penetrate to the moister undersoil. And the pine seedlings are far more drought-tolerant than are young spruces.

For these reasons spruces do not function well as pioneering trees in the open, but whitebark pine makes a very effective pioneer. Over the years the whitebark grove expands as its trees spread their crowns, and nutcrackers continue to bring in more seeds. The pine grove becomes a shadier, cooler place in the summer. In this changed environment, where larkspurs and bistorts, sulfur buckwheat and globeflower have given way to twinflower, woodland strawberry, Colorado columbine, and grouse whortleberry, spruces can finally take hold. They become established beneath the pines as they could not in the open. In effect, by planting pines, nutcrackers create spruce habitat.

The same thing happens in Mount Rainier National Park in the Cascades of Washington, where nutcracker-seeded whitebark pines growing in vast subalpine meadows act as "nurses" for mountain hemlock, eventually forming mixed-species tree islands. And in the Rocky Mountains, limber pine, another nutcracker-dependent pine, paves the way for Douglas-fir on dry outcrops where this less drought-resistant conifer has difficulty becoming established.

The pioneering role of whitebark pine is not restricted to meadows and cliff faces, but is often manifested on burned areas, ridgetops, and clearcuts. The Siberian stone pine functions in the same way, and also comes in on *shelkopriadniks*, forests catastrophically defoliated by the Siberian silkworm. And in the Alps, when Bergamask sheep were banned from pastures above the upper forest limit to prevent the spread of foot-and-mouth disease, groups of Swiss stone pine seedlings began to appear, regenerating groves that had been decimated centuries earlier.

Stephen Vander Wall and I studied the natural reforesting of a burn over ninety years old, 8,400 feet high in Utah's Raft River Mountains. The burned area had supported a mixed conifer forest of alpine fir, Douglas-fir, and limber pine which, judging from the abundance of standing snags and fallen timber, had been a reasonably dense stand. While alpine fir and Douglas-fir were becoming reestablished by wind-borne seed from downslope, limber pine seeds were arriving in nutcracker pouches. Apparently this had been going on since shortly after the fire, because one of the "new" limber pines was eighty-three years old. In the fall of 1979 the birds cached an estimated twelve thousand seeds per acre in the stony soil. Red Squirrels were absent from the mountainside, so the nutcrackers had no competitors in harvesting seeds from the pioneering limber pines that had already reached maturity on the burn. We estimated that 90.5 percent of the 1,461 cones on seventy monitored trees were removed before they opened. Seven percent of the cones remained on the trees long enough to open, at which point they were quickly relieved of their seeds by the nutcrackers. Cone beetles destroyed the rest of the cones. Most of the limber pine seeds we observed being cached on the burned site were harvested from the new generation of trees. The disparity between the large number of seeds cached and the small number of surviving pines was probably influenced by rates of seed theft by rodents and poor seedling survival on this harsh, windswept site. Perhaps the most dramatic examples of whitebark pine colonizing burns will be those that followed the great Yellowstone fires of 1988. Large areas of subalpine forests that are prime whitebark habitat went up in smoke that summer. And the following year—1989—saw a bumper crop of whitebark pine seeds throughout its Rocky Mountain range. Whatever stands result from that seed crop will of course be subject to the usual successional events. Just as in the open meadow of Squaw Basin, herbs, grasses and light-demanding shrubs will first flourish in the half-shade of the whitebark pine woodland. Then, more shade-tolerant conifers will come in, darkening the woodland to forest. Horned larks will give way to robins, robins to crossbills. Mosquitoes will rise up in clouds where once only brisk air cleansed sagebrush. A forest community will occupy once-open space, where nutcrackers pushed seeds into the ground, innocently catalyzing the building of an ecosystem.

The nutcracker plays its part as an ecosystem-builder—or "edificator," as the Russian ecologist N. F. Reimers (1953) put it—in a unique way, and with unique results. It does not scatter seeds over the landscape *randomly*, as does, for example, a robin, which having digested the pulp of a fruit, excretes in flight the hard seed within. A nutcracker deliberately *chooses* where each and every cached seed will go. We may not fully understand the rationale behind those choices, but that should not blind us to their deliberateness. One proof of this can be found in the clumped nature of whitebark pine groves and forests, another in the age structure of the forest.

Anyone visiting a grove of whitebark pine is immediately struck by the large number of trees that grow not as single, isolated individuals, but as members of clumps in which the stems make physical contact. As early as 1908, George B. Sudworth, the pioneering dendrologist of the United States Forest Service, observed that whitebark pines are "often in clusters of 3 to 7 trees, as if growing from the same root." Similar observations have since been made by numerous researchers of almost all the pines known or suspected to be dependent on nutcrackers for their regeneration.

Not surprisingly, these clumps have been attributed to multiple seedlings arising from nutcracker seed caches, and this has often been demonstrated by pulling up clump members and showing that each has its own root system, or by tracing the wood grain of clumped wind-thrown trunks into their own roots. In fact, when a conifer in nutcracker country is observed to form such clumps in high frequency, it is reasonable to hypothesize that its seeds are being cached by nutcrackers. The great majority of such conifers are pines with large wingless seeds, but as we will see in chapter 12, at least one western pine that has small winged seeds is also established by Clark's Nutcracker.

It is not always possible, however, to be certain that stem clumping is due to nutcracker caching. Rodent caches can also produce clumped stems, and if the leading shoot of a young pine is killed, several branches may turn and grow upward, eventually mimicking a clump. Pinyon pines are seldom clumped, mainly because their primary disperser, the Pinyon Jay, usually caches seeds singly, and perhaps also because competition for soil water on the semi-arid sites where they grow results in only one survivor. When winged-seeded conifers grow in a mixed forest with wingless-seeded nutcracker mutualists, their "clumpiness" is typically much lower than that of the mutualistic pine. For example, in a study made in Squaw Basin we found that of 1,270 whitebark pines, 47 percent were members of multi-stem clumps ranging in size from two to eight stems. But of 1,741 Engelmann spruces, only 1.5 percent were clump members, and the rare clumps never exceeded three stems. This is about what one would expect of a species whose wind-borne seeds land randomly, and occasionally germinate in close proximity to each other. Typical clumping frequencies of wingless-seeded pines are displayed in Table 9.1.

Some forest geneticists have been intrigued by the possibility that populations of pines cached by nutcrackers may exhibit a distinctive genetic structure or "architecture." One way of determining genetic structure—the distribution of genes within populations—is to analyze the frequency of occurrence of various molecular forms of enzymes in individual trees. Such a study of whitebark pine in the Canadian Rockies showed that members of a stem clump tended to be close relatives, while members of neighboring clumps were no more closely related to each other than were members of more distant clumps. This is quite different from the case of typical wind-dispersed conifers in which a distinct family structure occurs, with the trees forming patches of forest made up of individuals that are fairly closely related. It suggests that nutcrackers tend to put seeds from the same tree, or even the same cone, in caches that can then give rise to clumps made up of half-sibs—"half-brothers" and "half-sisters." But they also disperse seeds from the same tree more widely across the landscape than the wind disperses with winged conifer seeds. One probable consequence of the clumping habit

Table 9.1 Frequencies of multi-stem clumps among pines that have wingless seeds and are known to be nutcracker mutualists

Pine	Place and reference	Number of trees in sample	Percent of trees in multi-stem clumps	Stems per clump
Limber	Colorado (Woodmansee 1977)	470	56	2–15
	Utah (Lanner 1980)	186	68	2–4
	Utah (Lanner and Vander Wall 1980)	109	23	2–6
Whitebark	Wyoming (Lanner 1980)	1,270	47	2–8
Swiss stone	French Alps (Crocq 1978)	310	68	2–6
	French Alps (Lanner 1988)	167	89	2–11
	Swiss Alps (Mattes 1982)	345	31	2–>7
Japanese stone	Hokkaido, Japan (Saito 1983a)	69	94	2–19
	Hokkaido (Hayashida 1989)	69	97	2–11
Japanese white	Hokkaido (Hayashida 1989)	1,352	37	2–11

is that mating often occurs between half-sibs, because the pollen of one clump member is so easily blown onto the female conelets of another. One could predict that over the millenia, species in which this mating pattern occurs have developed a higher tolerance to inbreeding than have other pines, but little is known of these species' breeding systems.

The clumping of seedlings may have pronounced effects on their survival and growth. Claude Crocq has suggested that clumps handle snow loads better, apparently by forming a mini-canopy whose members are less likely to buckle under the weight of the snow than a lone seedling. Another "cooperative" effect that has been suggested for clump members is the sharing of resources through grafted roots; but root-grafting, and its resulting sharing of nutrients among trees, is ubiquitous among forest trees and does not require the close proximity of clumping for its occurrence.

The age structure of whitebark pine forests is also immediately obvious to the visitor. Large old trees grow shoulder-to-shoulder with young adults, with seedlings and saplings of all sizes and ages growing in the understory. New germinants may grow in the shade of battered veterans. This was evident in Karen Snethen's plots in Squaw Basin (Table 9.2), and also typifies the age distribution in limber pine groves throughout the Rockies and the

Table 9.2 Diameter-class distribution of trees in a single study plot in Squaw Basin, Wyoming, showing the all-aged nature of the whitebark pine population.

	Number of trees			
Size-class, DBH[a] (cm)	Whitebark pine	Englemann spruce	Subalpine fir	Lodgepole pine
Seedlings[b]	639	300	10	7
1–9	279	41	4	3
10–19	174	38	10	1
20–29	82	24	5	1
30–39	35	20	7	0
40–49	23	11	3	0
50–59	18	2	0	0
60–69	8	2	0	0
70–79	2	0	0	0
> 79	1	0	0	0
Total	1,261	438	39	12

Source: Data from Snethen (1980).
[a]DBH = diameter at 4.5 feet above the ground
[b]"Seedlings" are trees less than 4.5 feet tall, or taller trees of less than 1 cm DBH.

Great Basin. For example, in a small limber pine grove in Logan Canyon, Utah, thirty-one limber pines varied in age from twenty-four to approximately one thousand years.

This is clear evidence that Clark's Nutcrackers have repeatedly chosen to cache seeds on the same sites, decade after decade, century after century, far beyond the life expectancy of any single bird. In wind-dispersed pines, even those in the soft pine subgenus, a natural stand is much more uniform in its age structure, and seedlings are virtually excluded from mature forests. In these species, regeneration of the pine is highly unlikely to occur until a severe disturbance destroys the older trees.

The all-aged structure may be an attribute of all corvid-initiated pine forests. It is a prominent feature of Swiss stone pine forests in the Alps, and Siberian stone pine forests in Siberia. Limber pine groves are typically all-aged, and so are the woodlands of Colorado, singleleaf, and Mexican pinyons. David Ligon has called attention to the sense of place among Pinyon Jays who continue to cache pine nuts on sites from which the pinyon woodland has been eradicated by bulldozers in an effort to increase pasturage. European Nutcrackers are replacing long-gone stands of Swiss stone pine with new ones, on the very Alpine sites that were cleared and pastured centuries ago. A powerful instinct seems to be directing these seed-placement choices, perhaps as obsessive as the harvesting of pine nuts, and it is this instinct that triggers the process of forest succession.

Among the animals most directly affected by the successional changes set in motion by Clark's Nutcracker are Red Squirrels and Grizzly Bears. The same nutritional qualities of the whitebark pine nut that seem to obsess nutcrackers, and that enhance the survival of the pine on stressful sites, serve also to create a curious and intimate relationship between these mammals so disparate in size and habit.

CHAPTER 10

The Odd Couple

ED SQUIRRELS ARE SMALL AND LIVELY, forever running out on their aerial highways, making sprightly springing leaps between trees, loudly berating interlopers who enter their spiritedly defended little territories, comically playing the fool of the coniferous forest. Grizzly Bears are very large. They move on the ground with magisterial gait, sometimes plodding, sometimes pursuing in a burst of speed. They seldom vocalize loudly, and they roam over immense areas of wilderness. They play fearsome king to the squirrel's jester. Yet these mammals of contrasting physique and behavior are linked by the chain of events set in motion by Clark's Nutcracker. In brief, Red Squirrels hustle pine nuts for Grizzly Bears.

The Red Squirrel, *Tamiasciurus hudsonicus,* inhabits coniferous forests from Alaska south to Arizona, and from Quebec south along the spruce-covered spine of the Appalachians to South Carolina. They are opportunistic feeders during the summer, dining on a variety of mushrooms, which they dry on branches high in the forest canopy; on lichens stripped from rocks and tree-trunks; and on phloem scraped from the inside of tree bark. They also eat various fleshy fruits and berries from forest shrubs, including huckleberries, currants, and grouse whortleberries. But to face winter on equal terms, a Red Squirrel should have available a large cache of cones filled with nutritious seeds.

Recall that Red Squirrels—which are also known as pine squirrels, or chickarees—cut down the cones of whitebark pine and other conifers, and that they store most of them within "middens" in or on the ground. These middens, in the words of mammalogist Robert B. Finley, Jr.,

> are frequently 20 to 30 feet across, one to one and a half feet deep in the center, and carpet the ground to the exclusion of all living plants. The surface is usually littered with fresh cone scales and cores dropped by the squirrels. The material below is loose and damp, easily dug into with the bare hands. As one digs into the deeper deposits the material at lower levels is found to be older and more decomposed, forming a rich mulch in contact with the mineral soil. Large middens must be decades old and represent the accumulations of many successive generations of red squirrels. (Finley 1969, p. 234)

Occasionally, middens are made under water, by placing cones in cold mountain streams, or by cutting cones from trees leaning over the water along lakeshores and letting them fall several yards out from shore. Red Squirrels are proficient swimmers, so they have no difficulty retrieving cones placed in wet storage. There are benefits in storing cones where the wetness keeps the scales closed, the low temperature keeps the seeds fresh, and there is no odor to betray their presence.

The middens contain the squirrel's winter food stores, from which cones will be extracted, the scales gnawed away, and the seeds eaten, with scales, cores, and nutshells accreting to the debris of the midden. A midden may be stocked with as much as fifteen bushels of cones, and foresters have found them a rich source of seed for reforestation programs.

Red Squirrels are fiercely territorial, and make their cone middens on their territories, which usually center around a large tree. Since these territories are invariably in dense forest, middens are not found among the scattered whitebark pines of meadows like that of Squaw Basin, or in early-seral whitebark pine woodlands. But where late-seral whitebark pine is in mixture with spruce, fir, or lodgepole pine, middens are readily found.

And it is often Grizzly Bears who find them. Grizzlies (*Ursus arctos*) hibernate for about six months, beginning in October. Prior to hibernation they go through a hyperphaegic stage, a period of excessive eating during which a bear may add one hundred pounds to its body weight by growing a nine-inch-thick layer of fat. Good nutrition is a prerequisite to successful hibernation in all bears, but especially in pregnant sows. Implantation of blastocysts (fertilized ova) in the wall of the uterus is delayed for up to five months after mating, and thus occurs during the early stage of hibernation. If the new mother has ample fat reserves, tiny, finely-haired, blind cubs will be born six to eight weeks after implantation. If she does not have sufficient fat reserves, implantation will not occur. Cubs may weigh less than a pound at birth, but will grow rapidly on rich mother's milk that may contain 33 percent fat. Fat reserves also permit bears to remain in their grass-, duff-, or conifer-bough-lined dens for six or seven months without eating, drinking, or expelling liquid or solid wastes. Their metabolic mechanism allows them to compensate for the minor amounts of water vapor lost in exhaling by the breakdown of fat reserves, which liberates water. The urea formed during this process is broken down into carbon dioxide, which is respired, and nitrogen, which is incorporated into protein metabolism, thus preventing toxic uremia.

In the Yellowstone area the Grizzly's major pre-hibernation fat source is the whitebark pine nut. Viewed in the context of the total Grizzly diet, the centrality of nuts seems almost counter-intuitive, for a Grizzly is one of nature's great omnivores. This versatile animal—whose feeding tactics range from daintily picking berries with the lips and athletically pouncing on field mice to chewing down six-inch-thick trees—appears to crave fats, sugars, fiber, and proteins in all their organic forms. For appetizers, Grizzlies may eat ants, pocket gophers, and cutthroat trout; the salad course can include glacier lilies and horsetail, grass roots and spring beauty, elk thistle, fireweed, dandelion and clover, biscuitroot, yampa, and pondweed. For the main course: elk or deer—long dead or fresh-killed—or livestock do nicely; and few desserts surpass the fleshy fruits of grouse whortleberry, huckleberry, or soapberry, unless it be mushrooms, puffballs, or succulent army cutworm moths caught swarming on the scree. But when September and October crisp the meadows and turn the aspens, whitebark pine nuts are the discriminating Grizzly's first choice.

Grizzlies begin raiding middens in August, and continue through October. A bear encountering a midden plows it up with its heavily clawed paws. If the squirrel has pushed cones into the soil beneath the midden debris as well as within the debris itself, the great furry bulldozer simply tears up the soil. While many Grizzly excavations are apparently exploratory, and involve less than two cubic meters of material, a bear will occasionally move thirty cubic meters of soil and midden debris—about two large dump-truck loads—in its search for pine-nuts. Visitors to these forests in the fall should therefore beware the scent of fresh-turned earth, and consider the benefits of a hasty retreat.

What happens next has been graphically described in a memoir written by septuagenarian Vern E. "Bud" Cheff, Jr., a native of the Flathead Lake, Montana area, who frequently observed Grizzlies in the 1920s and afterwards in the Mission Mountains:

> They hold the cone between their paws and going right around the cone with their teeth, take all the shell off right next to the nuts, spit

> the shell out and then go around one more time, this time taking all the nuts off in one pass. It is great to watch as it takes only a few seconds to each cone. In the early part of the nut season a lot of the cones have pitch on them so that the bear's paws and forearms and nose would be covered with pitch, but by hibernation time most of it would be off. The later the season, the drier the pitch becomes on the cones. At this time of the year the grizzly droppings were nearly 100% pine nut. (Cheff 1984, p. 6)

By "shell," Cheff is apparently referring to cone scales. Kate Kendall, who has studied the biology of Grizzlies in Yellowstone and Glacier National Parks (Kendall 1983), has described another variation. She was curious to know how bears were able to consume whitebark pine nuts without consuming the cone scales, so she visited the Boise, Idaho zoo to feed captive Grizzly and Black Bears. "Despite ten years of captivity," Kate told me, "the Grizzlies knew exactly what to do." They ignored shelled nuts Kate offered them, but crushed the cones with their paws or broke them in their jaws. The cone debris was spread out with paws or muzzle and the nuts delicately licked up, while cone scales were expelled out the side of the mouth. Even moldy cones from deep in a midden were accepted. The bear droppings from these trials contained no cone debris.

Kate Kendall has pointed out that Grizzly Bears dig up middens in the spring *after* having hibernated, as well as in the fall. At those times they raid cones of the previous year's crop. For reasons known only to the bears, they never seem to take all of the cones, but rather leave at least a few behind. In spring there is usually heavy snow cover at the high elevations where bears den, and they may plow through snow five feet deep in order to raid a midden of whitebark pine cones. Whether they locate the middens by scent, or by visual cues, like the nutcracker, is not known.

Professor Charles Jonkel of the University of Montana has observed Grizzly Bears pulling whitebark pine cones from the low branches of small trees growing in the open, and becoming heavily smeared with pitch in the process. The land-bound Grizzly cannot climb, so nearly all of its pine nuts must be stolen from Red Squirrels. Black Bears, however, are expert tree-climbers, and sometimes climb into whitebark pines, where they forage pine nuts directly from the trees instead of relying on Red Squirrels as middlemen.

The importance of the whitebark pine cones that Yellowstone Grizzlies find in middens cannot be overstated. Consider for example, the research findings of David Mattson and his colleagues in Yellowstone National Park. In years when whitebark pine was "masting"—producing very heavy cone crops—pre-hibernation Grizzlies appeared to feed during August through

October almost entirely on pine nuts, and they tended to concentrate their foraging in the mixed subalpine forests where squirrel middens were most concentrated. Because these were usually in relatively remote areas, large nut crops kept the Grizzlies where they were unlikely to encounter humans. But when the nut crop was small or entirely lacking, Grizzlies were twice as likely to turn up within three miles of roads and five miles from developed sites. As a result, six times as many bears had to be trapped and transplanted in order to safeguard hikers, campers, and tourists using the back country and campgrounds. Some bears keep coming back, despite being transplanted, are considered dangerous, and are killed. Thus two to three times as many bear deaths occurred during years of poor nut crops.

Obviously then, it is in the interest of both Grizzly Bears and humans that whitebark pines in remote areas prosper, and bring forth great treasures of pine nuts. And it is equally important that the Red Squirrel prosper, because this small mammal forms the bridge that connects the Grizzly Bear with its winter sustenance.

CHAPTER 11

Pine Nuts and People

N THE FALL OF 1983 AMERICAN FOREST entomologist William Mattson was riding the Trans-Siberian Railway between Irkutsk and Moscow. Despite his fancied familiarity with Russian custom, he was astonished to see his fellow passengers reach into their bags or pockets, extract egg-shaped pine cones, and shuck the scales one by one as they threw the nuts into eager mouths. The train's conductor came down the aisle, reached into a cloth sack, and distributed pine nuts to passengers who wished a snack with their tea. As a biologist, Mattson should not have been so taken by surprise. After all, *Homo sapiens* is a resourceful animal, ever alert for nutritious foods, and has long fed on pine nuts. Among the soft pines, these include seeds of the pinyon pines of subsection *Cembroides*, the two Asian species of subsection *Gerardianae*, several of the *Strobi*, and all five species of *Cembrae*, the stone pines.

In another book I dealt at length with the central importance of pinyon nuts to the native peoples of the American Southwest (Figure 11.1) and Mexico (Lanner 1981). Here I will focus on the stone pines, whose story is equally fascinating but less well known. Like the pinyons, the stone pines inhabit stressful environments, where human survival is a constant challenge, but while pinyons grow on the desert margin, most stone pines grow high on the slopes of snowy mountains or in bleak subarctic milieus.

Whitebark pine nuts were well known to the Interior Salish tribes from Montana to British Columbia—the Lilooet, Nlaka' pamux, Shuswap, Okanagan-Colville, Chilcotin, Kootenay, and Flathead. Charles Sprague Sar-

FIG. 11.1 Artist's conception of an eighteenth-century Shoshoni pine-nut camp in eastern Nevada. The women store pinyon cones in a stone ring while a man beats cones from a singleleaf pinyon tree. Sketch by Rose Milovich. Courtesy of Steven Simms, Utah State University Anthropology Museum.

gent (1897) wrote that "the sweet seeds were gathered and eaten by the Indians, although Clark's Crow, which tears the cones to pieces before they are ripe in order to devour the seeds, left them only scanty harvests." According to ethnobotanist Nancy Turner of the Royal British Columbia Museum at Victoria, the Thompson and Lilooet Interior Salish referred to whitebark pine in their language as the pine-nut plant. Nuts were collected by women and children while the men were off hunting in late summer and fall. Children climbed into tree crowns to pick the cones, break or cut off cone-laden branches, or shake the limbs until the loosened cone scales and seeds rained down. The nuts were sun-dried or roasted, and then stored in cloth bags. They were often eaten mixed with "saskatoons," the small purplish-black pomes of the western serviceberry (*Amelanchier alnifolia*), or they were taken in a mush made with mountain-goat fat or water.

The Flatheads of Montana did much the same, according to Bud Cheff, who as a boy in the 1920s accompanied Indian friends into the Mission Mountains. Groups of up to twenty family members and friends camped in the forest, where they hunted, fished, prospected, dug roots, picked berries, and plucked cones from the pine-nut trees while squawking nutcrackers flocked overhead. The cones were brought down to the reservation in burlap sacks slung over packhorses. The nuts were extracted from the cones and roasted, and doled out over the winter as a snack.

Carling Malouf, a University of Montana anthropologist, recalls that pine nuts were sometimes processed in the mountains. The seeds were removed from the cones, roasted, shelled on a flat rock with a stone pestle, and sacked, in the high country. This was women's work, he stresses, part of the annual round of food-gathering activity. All sources agree that whitebark pine nuts were never the staple food that pinyon nuts were in the Southwest. But they were recognized as a rich snack food, and the pine-nut harvest was an important social occasion.

Arvennüsse—the nuts of Swiss stone pine—were an important traditional holiday snack in the Alps. They were especially esteemed, it is said, in helping to while away the long Alpine winters. Large quantities were consumed locally and they were exported to the markets of Augsburg and Munich as well. Pine nuts were the essential ingredient of the famous nut-tortes of the Engadin Valley, but their gathering was not always a pretty sight. The Swiss historian Pol has described how, in the seventeenth century, forest authorities would allow cone-gathering during one day prior to a feast-day, at which *einer feindlichen Schaar*—a hostile troop—of old and young stormed the designated harvest area where, not satisfied merely to pick fallen cones from the ground, they proceeded to climb into the crowns of second-growth stone pines to knock down cones with their sticks, breaking branches in the process, causing the wounded trees to bleed pitch, and creating an uninhabitable wasteland. According to Switzerland's stone pine historian M. Rikli (1909), peasants even felled stone pines to harvest the cones, a level of vandalism no longer seen in the carefully-tended forests of that country.

The westernmost of the scattered islands of Swiss stone pine in the Carpathians is found in the High Tatra Mountains of southern Poland and northeastern Slovakia. Here both countries have set aside small national parks to preserve the remnant stone pine stands that have survived centuries of degradation visited upon them by shepherds and woodcutters. In the twelfth and thirteenth centuries German colonists who settled in the Spisz region at the foot of the Tatras called stone pine by the names *Limbaum*, *Linbaum*, and *Leimbaum*. By the sixteenth century, Poles used *linba* or *limba*. Today *limba* is the recognized Polish name, and is used as well in Czech and Slovak.

Stone pine nuts were gathered in the Tatras, and were known to have been sold as a foodstuff in local markets in the nineteenth century, but little has been written on the subject. The nuts were reputed to have unusual medicinal properties, as shown by the following prescription from a seventeenth-century Polish manuscript: "If someone's mind deteriorates, stone-pine nuts . . . should be pounded into a powder together with laurel oil, and having made the mixture use it to anoint the temples above the ears." "Stone-pine oil," distilled from shoots of the tree, was also used against various ailments, and stone pine wood ashes were taken in wine.

The great utility of the nut produced by the Siberian stone pine, or *kedr*, lies first in its own good qualities, and second, in the ubiquity of the tree that bears it. In most areas the fat content of the nuts varies from 60 to over 70 percent, and the protein content is about 19 percent. As for ubiquity, *kedr* occupies about 100 million acres of the world's largest coniferous forest. Vast amounts of nuts become available in good seed years. This highly nutritious food did not go unnoticed by the numerous ethnic groups that populated Siberia. Among the peoples that gathered cedar-pine nuts were the Evenki, the Tatars, the Todzhans or "Reindeer Tuvans," the Tofalars, and the Shors.

The Evenki started collecting pine nuts late in the summer when the shells were still soft, and chewed them up, shell and all. Later, nuts were baked in hot ashes, pounded, and mixed in boiling water to form a gruel, or they were eaten mixed with meat. The Tatars had a long tradition of procuring pine nuts in which they divided the forest into family gathering zones for the harvest. Todzhan women collected pine nuts, which they called *kuzuk*, and traded them to Russian merchants for such necessities of taiga life as rifles and clothing. The Russians who had moved east and settled in Siberia acquired the habit of chewing *senka*—lumps of larch resin—and pine nuts. Levin and Potapov (1964) write that the pine nuts "were chewed in large mouthfuls and in silence, particularly when people went visiting. This is jokingly called the Siberian conversation."

The Shors, a small Turkic-speaking people, had traditionally gathered pine nuts in relatively small quantities, using them less as a staple food than as a supplement. But when Russian traders came into the taiga, the Shors made nuts a cash crop. Family groups would go out into the forest in August of a seed year, just as the nuts were maturing. They set up camps of huts to live in, and silos for nut storage (Figure 11.2). By the end of the 1800s a majority of Shor families in some areas was involved in the nut trade. But not all prospered. Those in financial straits who needed quick cash sold their wares at a low price to the Russian buyers, or used them as payments on the previous year's debts. Dealers often exploited the nut-gatherers, paying them with equal amounts of barley, or smaller amounts of grain. It was only the more able Shors who took their harvest back to their villages, extracted and stored the nuts, then sold them later when the prices rose. Some of these entrepre-

FIG. 11.2 A bear-proof bin for storing cones in the forest. (Lebedeva and Saf'yanova 1979)

neurs dealt with the Russian dealers, but others traveled the steppes to do business with tribespeople who lived far from the wild nut groves of the taiga.

Despite the excellence of the Siberian stone pine nut as a food, and notwithstanding its almost boundless supply, gathering large quantities of nuts was never easy. Animal competition was one reason. Nutcrackers pecked cones from the trees; squirrels cut them down, carried them away, and buried the seeds; chipmunks cached great numbers of cones. According to Russian foresters, nutcrackers are attracted by the activity of people collecting cones, and quickly fly in to offer competition. Severe frosts often reduced crops in an early stage, and cone moths frequently destroyed large numbers of young cones. More difficulty was created by the sheer size of the nut-bearing trees, and the height of their crowns. The *kedr* is a respectable forest tree, often reaching heights of one hundred feet or more, and ages of up to six hundred years. Forest-grown trees have narrow crowns, hemmed in by those of neighboring trees. Cones develop almost entirely in sunlit parts of the crown, which, in a forest tree, means up on top. Big trees bear many more cones than small trees, but their cones may be all but impossible to reach. Smaller trees are easier to harvest from, but they yield fewer cones. Despite reports that young boys sometimes climbed the trees, the standard method of getting cones was to gather fallen cones—"windfalls"—before the rodents got them, or by the use of the *tokpak* (Shor), or *kolot* (Russian). This is a great mallet or hammer made from a short log fitted with a six-foot-long handle. The base of the handle is set on the ground a meter or so from the tree, and the malleteer—sometimes with assistance from companions—rams the head against the selected tree-trunk (Figure 11.3). If all goes well, a shower of cones results, because unlike whitebark pine cones, those of the Siberian species are only loosely attached to the branchlets. Too many vigorous blows, however, might loosen or tear the bark, especially on a thin-barked young tree, so the method is not without risk to the trees. In fact, according to Konstantin Krutovskii, use of the *kolot* has been illegal for

FIG. 11.3 Workers hammering a Siberian stone pine with the *kolot* to shake cones loose. (Lebedeva and Saf'yanova 1979)

many years. The cones are then gathered up and spread on a *mashinka*, a washboard-like device, on which they are cracked apart to release the nuts (Figure 11.4). The nuts are winnowed from the cone debris, and sometimes roasted over an open fire in a large tray suspended from a long boom by a rope. As recently as the 1960s, nuts were hauled out of the forests of the Transbaikal in sacks slung over a horse's back, or in two-wheeled horse-drawn carts. Forester A. M. Kozhevnikov (1963) estimated that procurement of one ton of stone pine kernels by four men from a forest forty-five kilometers from the nearest settlement would require fifty-two man-days and forty horse-days. At the fixed price of kernels, the venture would only become profitable when the harvest exceeded about thirty kilos per hectare.

In the 1960s ambitious plans were made for utilizing the vast Siberian stone pine resource by the manufacture of wood products, large-scale production of cooking oil from the nuts, and perhaps resuming the nut exports first begun in the fifteenth century by Czar Ivan IV, "the Terrible." But while production rose to over a million tons in 1973, it appears these plans have never come to fruition. The sheer size of the country, the difficulties of access, the uncertainties of nut yields, and the primitiveness of technology have made the dream of turning Siberia into a giant nut orchard as elusive as Trofim Lysenko's fevered vision of making it a vast citrus grove. In addition, heavy overcutting of stone pine during the last century has decimated forests that formerly surrounded such commercial centers as Tomsk, Irkutsk, and

FIG. 11.4 Siberian stone pine cones are cracked open and the nuts removed by rolling them across a *mashinka* with a block of wood. (Lebedeva and Saf'yanova, 1979)

Krasnoyarsk. At present, it appears that Siberian pine nuts are used mainly as a snack food, and their shells are used to a limited degree to flavor vodka.

The habitat of Japanese stone pine—*stlannik* to the Russians in whose territory most of it is found—is graphically conveyed in a short passage from the journal of the mariner Martin Sauer. Sauer was a fellow Briton of Commodore Joseph Billings, who was engaged by Her Imperial Majesty Catherine the Second, Empress of all the Russias, to map the coasts of northeastern Siberia and Alaska late in the eighteenth century. Sauer was proceeding overland with several companions from Moscow to Aldan in eastern Siberia on Wednesday, June 24, 1786. Sore of foot, they crossed a bog and then ascended a mountain called Unechan, about 118 miles from Aldan. On top "we had a shower of snow, and were quite benumbed with cold. We crept under the Pinus Cembra, made a fire, heated some water with brandy, and refreshed all hands." Sauer later said of this shrubby pine:

> The creeping cedar, or pinus cembra, produces a considerable quantity of seeds or nuts in cones, like the common pine; but they ripen only the second year. Immense numbers are collected by the inhabitants; sometimes a considerable quantity are found in the squirrels' nests in hollow trees; in fact, they are the chief food of squirrels and mice. A very pellucid and sweet oil is extracted from these seeds. (Sauer 1802, p. 91)

Sauer was at this time among the Yakuts. His reference to *Pinus cembra* stems from the old concept that Swiss, Siberian, and Japanese stone pines are of the same species, *Pinus cembra* as originally described by Linnaeus. In February 1787, when Sauer and his shipmates were building ships on the coast of the Okhotsk Sea, "the scurvy gained ground among our people, affecting their joints, particularly the legs. A decoction of the *Pinus cembra* was used . . . and with success." A century and a half later, Soviet medical researchers prescribed the same remedy for this scourge of the north. According to Siberia specialist Gail A. Fondahl of Stanford University, it was also administered to prisoners of the *gulags* in the Kolyma region who were forced to do hard labor in local mines during the Stalin era.

Soviet scientists in the 1930s remarked on the high oil content of *stlannik* nuts, about 50 to 63 percent of kernel weight, and speculated on its potential use in medicine, pharmacology, and the cosmetics industry. They noted that the gathering of great volumes of nuts was facilitated by the pine's dwarfish habit, which permitted cones to be picked directly from the treetops, much as one would pick huckleberries from the tops of high bushes. But *stlannik* nuts have never attained the economic importance of Siberian stone pine nuts. One report states that a delicious and nutritious drink—"nut milk"—

can be prepared from *stlannik* nuts, but no recipe is provided. And according to a Russian observer, the Evenki of the Okhotsk region were fond of young green cones of the creeping pine, roasted in the ashes of their fires, and eaten "like a potato."

In Korea the nuts of the Korean stone pine have long been reputed to have therapeutic value. When Harry and Sue Hutchins perched high on a fire tower in the fall of 1992 to observe the interaction of nutcrackers and Korean stone pines in China's Little Quingan Mountains, their reverie was shattered by noisy mobs of local residents who invaded the woods in a competitive scramble to gather up fallen cones. Many climbed recklessly into the trees to shake cones loose, or even to break off cone-laden limbs. Occasional overachievers fell out of the trees and incurred serious injury, and had to be carried from the fray. In fact, seven deaths occurred in the area in 1992. Were these trees providing badly needed foodstuff to an impoverished peasantry? Hardly. Frenzied local farmers were doubling their annual income by collecting a free cash crop for export to Korea and Japan. Like Pol and Rikli before him, Hutchins became exasperated by the conduct of the crowds below. The nutcrackers' thoughts were not recorded.

CHAPTER 12

Deviations

IOLOGISTS ARE OFTEN PERCEIVED AS PEOPLE who give complex answers to simple questions. But biology is almost never simple.

Take for example the preference of Clark's Nutcracker for large, wingless pine seeds. At first glance, the preference makes perfect sense: large seeds have more nutrition and energy resources than smaller ones, and if all other things are equal, the bird increases its foraging efficiency by harvesting fewer and larger seeds. Wingless seeds should also be a wiser choice than winged ones. For one thing, a winged seed is unlikely to stay put when the cone opens, but will soon flutter to the ground. A wingless seed, on the other hand, is far more likely to remain in the cone where the nutcracker can find it. And before a winged seed is ingested, the wing must be removed. On the whole, this simple pattern is observed, but there are exceptions. It has already been mentioned that Clark's Nutcrackers feed on the winged but relatively large seeds of ponderosa and Jeffrey pines, and those of Douglas-fir. But Diana Tomback once saw these corvids taking seeds from lodgepole pine cones, and lodgepole pine seeds are among the smallest in the genus, as well as being winged. A lodgepole seed may have only one-fiftieth the heft of a whitebark pine seed.

Somewhat less extreme, but of more ecological significance, is the feeding of Clark's Nutcrackers on the seeds of Great Basin bristlecone pine. These are about one-twentieth the weight of singleleaf pinyon, which nutcrackers commonly harvest in this land of dry basins and mountain islands between the Sierra Nevada and the Rockies. Great Basin bristlecone is the pine that is world-famous as the oldest living thing, several trees having attained ages of over four thousand years. Its name usually invokes images of contorted timberline trees growing on sterile cliffs at ten to eleven thousand feet above sea level, barely clutching to life as they are buffeted by incessant gales. Indeed, the best-known stands of these trees do grow at high elevations in rigorous climates, as exemplified by those of California's White Mountains and Nevada's Great Basin National Park. But the species also grows as low as eight thousand feet in southern Utah and Nevada, where tall, straight specimens mingle with ponderosa pine and white fir.

In these stands the bristlecones grow as single-stemmed trees scattered among their neighbors, but with increasing elevation the frequency of clumped stems increases. Thus in a canyon of Mount Charleston, Nevada, at 8,300 feet, only 13 percent of the bristlecones were in clumps, but at 8,700 feet the figure rose to 51 percent, and at 9,400 feet it was 75 percent (Table 12.1). In the White Mountains, clumping characterized 46 percent of the trees in the Schulman Grove (10,200 feet), but 69 percent in the Patriarch Grove (11,300 feet). Nutcrackers were common in most of the fifteen locations where I made these and similar observations, as were cones that had been riddled and shredded in the nutcracker fashion while still closed. In thirteen of the locations, limber pines or pinyon pines were present. It appears to me that the nutcrackers are attracted to the bristlecones because of their relatively reliable seed production, for there is often a supply of these small seeds in years when the limber and pinyon pines are bare of cones. Feeding on the small bristlecone seeds is tedious, but it may be preferable to the uncertainties of an eruptive migration. Most of the surviving bristlecones at higher elevations seem to have become established from nutcracker seed caches, perhaps because wind-dispersed seeds on these high, dry summits would suffer from dehydration. If I am right, then the bristlecone's success as a high-elevation species is due to the Clark's Nutcracker's habit of making seed caches there.

Another relationship that *should* be pretty simple is that between Siberian stone pines and the mammals associated with it. The Siberian stone pine is found where there are nutcrackers, squirrels, bears, and chipmunks, just like the whitebark pine. But the differences between these two species of pine may be as significant as the similarities.

Our Grizzly Bear is regarded as a subspecies of the species *Ursus arctos*, the Brown Bear, which is patchily distributed in the Old World from the Pyrenees to Kamchatka, and in the New World from Alaska to California and from the Northwest Territories to the northern Rockies. Brown Bears once lived in the Alps and Carpathians where Swiss stone pine grows, and their range across Siberia and in the Far East corresponds closely with that

Table 12.1 Stem clumping, probably due to seed caching by Clark's Nutcracker, of Great Basin bristlecone pine (*Pinus longaeva*) stands in California, Utah, and Nevada.

Location	Elevation (feet)	Trees sampled	Percent of trees in stem clumps
White Mountains, Calif.			
Schulman Grove	10,200	76	46
Patriarch Grove	11,300	122	69
Snake Range, Nev.			
Mt. Washington 1	11,150	96	71
Mt. Washington 2	11,085	77	51
Frisco Peak, Utah	9,640	148	59
Huntington Creek Plat., Utah	9,020	352	71
Markagunt Plateau, Utah			
Cedar Breaks National Monument	10,430	171	76
Bristlecone Trail	9,805	69	80
Mammoth Creek			
above escarpment	8,445	63	25
below escarpment	8,365	128	63
Birch Spring Knoll Road			
above escarpment	8,345	79	39
below escarpment	8,300	156	58
Mt. Charleston, Nev.			
transect, lower	8,300	120	13
transect, middle	8,700	172	51
transect, upper	9,400	511	75

Source: Data from Lanner (1988).

of the stone pines that grow there. In fact, most of the Brown Bears still surviving in the Old World live in the Sayan and Altai Mountains, the Baikal region, and the Sikhote Alin, all of which have great areas of Siberian or Japanese stone pines. North America has Clark's Nutcracker, while Europe and Asia harbor the more variable and widespread Eurasian Nutcracker. Whitebark pine forests of the Rockies are inhabited by the Red Squirrel, *Tamiasciurus hudsonicus*, or its close cousin the Douglas Squirrel (*T. douglasi*) in the Cascades and Sierra Nevada. In the Eurasian stone pine forests dwells the *other* Red Squirrel, *Sciurus vulgaris*. It would be satisfying to report that in all stone pine areas worldwide, nutcrackers plant, squirrels harvest and store, and bears reap. But it is not quite that simple.

The Red Squirrel of North America and the Red Squirrel of Eurasia live in similar habitats with similar climates. Yet, according to British mammalogist John Gurnell, "they exhibit very different social organizations which seem to be linked to different seed hoarding strategies." *Our* Red Squirrel is fiercely territorial, and defends its large caches, or middens. Such concentrations of stored food are called "larder hoards," and they are probably defended because they represent a large investment of energy expended, and are therefore worth it.

The Eurasian Red Squirrel, however, is not strictly territorial, but lives in home ranges that overlap those of its neighbors. It disperses its stored food in numerous small, scattered caches called "scatter hoards." These, like the small caches of nutcrackers, are not worth fighting over. But if scatter-hoarded pine nuts do not provide a huge concentration of calories in one place, they cannot promise much for pre-hibernation Brown Bears. So how does the Eurasian bear benefit from the produce of the stone pine? Enter Burunduk.

Burunduk, the Siberian Chipmunk (*Eutamias sibiricus*), appears to be the one rodent that forges a strong link between Siberian stone pine and the Brown Bear. These little animals store a variety of wild seeds, including those of larch, maple, linden, hazelnut, bird cherry, raspberry, and sedge, as well as stone pine nuts. In agricultural areas they hoard corn, wheat, barley, buckwheat, and sunflower seeds. While no data seem to be available on the size of their stone pine seed hoards, caches of up to thirteen pounds of unspecified foods have been reported, and several Russian researchers have reported that these caches are excavated by Siberian Brown Bears.

This is in marked contrast to the behavior of chipmunks observed by Harry Hutchins in our whitebark pine study sites in Squaw Basin, Wyoming. Hutchins found they had little interest in pine nuts, and he could find neither nuts nor cones in the chipmunk burrows he excavated. Thus Hutchins contested the older studies that claimed, without sufficient evidence, that chipmunks are an important agent of conifer regeneration in the Rockies. But, apparently, not all western chipmunks are equal. From 1988 to 1990, Stephen Vander Wall conducted a careful study of the interactions of a Jeffrey pine seed crop and a suite of scatter-hoarding rodents in the Sierra Nevada. Jeffrey pine has a prominent seed wing, and has always been assumed to be dispersed by wind, but the seed kernel is large and is known to be taken at least occasionally by nutcrackers. The most common rodent species in the area was the Yellow Pine Chipmunk, but there were also Lodgepole Pine Chipmunks, Long-eared Chipmunks, Deer Mice, and two species of ground squirrel, all of which are known scatter-hoarders. Nutcrackers and Steller's Jays were also common, and when a very large seed crop was deposited on the ground, Black Bears licked up seeds in large numbers.

Vander Wall's field observations and experiments with captive animals showed that the Yellow Pine Chipmunk (*Tamias amoenus*) was the most important cacher of Jeffrey pine seeds. By concentrating fallen seeds into subsoil caches, the chipmunk not only dispersed them further from the parent tree, but also placed them in a seedbed favorable for germination and establishment. Thus the Yellow Pine Chipmunks behaved much like nutcrackers, but with shorter dispersal distances. Vander Wall's results show that dispersal of winged seeds is sometimes a two-stage process, with rodents playing an important role as seedling establishers. Whether other pines will prove to benefit so significantly from rodent activity awaits further careful research. It should be noted that Vander Wall's results in no way weaken Hutchins's conclusions that chipmunks play no role in whitebark pine establishment. As a second-stage disperser of Jeffrey pine seeds, a chipmunk need only stay on the ground, and it deals not with cones but only with seeds. To effectively disperse whitebark pine seeds, however, a chipmunk would have to climb into the tree crowns and deal with cones that do not release their seeds. This may be a challenge our western chipmunks are not up to.

Another example of "deviant" behavior is that of a squirrel mimicking a nutcracker—or at least having the same ecological function. This is sometimes the case with seed dispersal and seedling establishment in Korean stone pine, *Pinus koraiensis*. Until recently Korean stone pine had received little attention from researchers, so details of how its seeds are dispersed were entirely lacking. But then a compelling picture emerged of a pine-squirrel symbiosis. During the 1980s, careful studies were made by Masami Miyaki and Mitsuhiro Hayashida in planted stands of the pine on the Japanese island of Hokkaido. The fauna of the area was believed similar to that of areas in China and Korea where the pine is native. Though Miyaki and Hayashida worked independently, many of their findings are remarkably similar and mutually supportive. A synthesis of their results follows.

The cones of Korean stone pine are the largest of the *Cembrae*—growing to nearly eight inches long and weighing almost a pound when the seeds mature in late August (Figure 2.6). Their seeds are also the largest of the *Cembrae*, and have the thickest seed coats, about 0.88 millimeters thick, twice that of Japanese stone pine. In late August, *Sciurus vulgaris orientis*, the Far Eastern variety of the Eurasian Red Squirrel, starts to harvest the moist, pulpy cones, which are weakly attached and readily separate from the branchlet. The squirrel drops them on the ground, runs down the tree trunk, and strips the thick apophyses from their cone scales. This last act reduces the cone's weight by half, making the burden barely transportable by the squirrel, which frequently drops its cone as it carries it away. When a cone is dropped the seeds remain within. After carrying a cone for awhile, the squirrel puts it down, removes a seed, and buries it about three centimeters deep by holding it in its mouth while pressing it into loosened soil. The squirrel goes back to the cone, removes another seed, and buries it in the same spot, making caches of two to five seeds.

"After making several caches in one site," writes Hayashida, "the squirrel transported the cone to a new site and repeated the procedure." Both Hayashida and Miyaki documented successful establishment of young pines well outside the parent plantations. In Miyaki's study sites, pine seedlings became established up to six hundred meters from the source of seed. Hayashida found 98 percent of seedlings to be within one kilometer, but one was 1.8 kilometers from the nearest possible parent tree. Both investigators agreed that the Red Squirrel was the most important disperser of the Korean stone pine, and that nutcrackers had no apparent role in moving seeds.

Blissfully unaware of these Japanese studies, Harry and Sue Hutchins made preparations to visit northeastern China in the fall of 1992. Their purpose: to determine whether Eurasian Nutcrackers disperse Korean stone pine seeds. By late September they had spent a week shivering atop a thirty-five-meter fire tower in the lush mixed conifer-broadleaf forest of Heilongjiang Province. Despite the onset of the fall color season, Harry and Sue felt little cheer. The sky was overcast, the temperature near freezing, and their quarters unheated. Worse yet, the insistence of their Chinese hosts at Harbin's Northeastern Forestry University that only squirrels dispersed Korean stone pine seeds had nearly persuaded Harry and Sue that they were committing the boondoggle of their young lives. But their faith in nutcrackers was soon vindicated. Within a week they concluded that nutcrackers were indeed important, both in the dispersal of Korean stone pine seeds, and in the establishment of seedlings. Nutcrackers removed the cone scales, harvested and pouched seeds—at least sixty per filled pouch—and flew with them up to at least four kilometers. The Hutchinses observed nutcrackers making seed caches in the forest litter, and they dug up the caches to prove that pine nuts were contained therein. Harry and Sue found young seedlings growing in the tell-tale clumps that result from multiple germinations in a nutcracker's cache. They observed Japanese Grosbeaks, hawfinches, various woodpeckers, and nuthatches stealing pine nuts from cones opened up by the nutcrackers. There was little squirrel activity, probably because heavy trapping in the 1980s had decimated squirrel populations.

Why do the Hutchins's observations differ so starkly from those of Miyaki and Hayashida? Probably because they were made in the pine's natural habitat. The northeastern Chinese forest was part of a vast area of native pine, larch, spruce, fir, ash, poplar, maple, birch, and oak woods. There were about fifteen thousand acres of old-growth stone pine. The Japanese studies, on the other hand, were conducted in man-made forests, plantations of stone pine in which nutcrackers were seldom seen. Whether Red Squirrels are important dispersers or establishers of the pine under natural conditions remains to be learned. Harry and Sue Hutchins's findings clearly show, however, that the stone pine-nutcracker mutualism extends to all of the stone pines, and that a single evolutionary scenario should be able to embrace the relationships of all the stone pines and nutcrackers.

Having considered the activities of a squirrel that acts like a nutcracker, it is only fitting to consider a nutcracker that acts like a squirrel. The Tian Shan Nutcracker mentioned in chapter 7 inhabits pure forests of the Tian Shan spruce in what was once Soviet Central Asia, and is now the republic of Kyrgyzstan. According to a 1936 Russian report by S. V. Kirikov, the nutcracker, which appears to depend on a diet of spruce seeds, grasps a spruce cone in its bill and whips it back and forth until it comes loose. If this does not work the nutcracker "clutches the cone in the toes and hanging head downward tears it off." The nutcracker hammers some of the cones apart, to gain access to the seeds. Other cones are carried off some distance. Kirikov relates:

> Digging under the spruces in the moss near the place where [the nutcracker] had been swooping down, on its side I found a young cone of the year hidden in the moss, and then another, and there a third; there was a cluster of seven, like pepper mushrooms. (Kirikov 1936)

Despite his insistence that nutcrackers cause seedlings of the spruce to become established, Kirikov offers no supporting evidence. It is not at all obvious that the caching of spruce cones would result in the germination of their seeds. More careful observations are needed of the relationship between these spruces and the nutcrackers that make middens like those of squirrels.

Deviant behavior is not the sole province of animals. Plants, too, can behave in unconforming ways, even stone pines, and by doing so can complicate what first appeared to be a simple and straightforward mutualism.

Take for example the issue of seed dormancy. Some pines' seeds can germinate virtually right off the tree, in the autumn of their maturation. The seeds of most species, however, are shed in the fall, and overwinter in or on the soil, presumably because they need the moist chilling treatment to prepare them for spring germination. But the pines of subsection *Cembrae* follow their own rules.

All the stone pines are reputed to have seeds that can remain dormant for several years after they mature. Forest Service researcher Ward McCaughey has documented a higher rate of whitebark pine seeds germinating as late as the third, and even the fourth year, than in the spring following dispersal. The results of these controlled experiments were supported by field data in areas burned over by the great Yellowstone fires of 1988: whitebark pine seeds from a mast crop cached in the fall of 1989 were observed germinating in 1990, 1991, and 1992.

The easy explanation of this staggered germination habit is to label it an adaptation of whitebark pine to life in a variable environment. Whitebark pine, according to this view, is hedging its bets by spreading out over several years the sprouting of seeds from a single year's crop. So if the year following seed dispersal turns out to be hot and dry, or cold and soggy, or in some other way inimical to tender young seedlings, well, let's try some more next year and see how we do. And the year after, and maybe the year after that. This is whitebark pine's way of spreading its risk.

But that explanation has difficulties. For one thing, it is obvious that a seed left longer in the ground is vulnerable to rodent predation that much longer. And if this strategy works so well for whitebark pine, why not for other pines as well? All pines, as well as other conifers, grow in variable environments. But they do not spread out germination of a seed cohort over several years.

Another explanation could be based on nutcracker behavior. Recall that nutcrackers seem to cache all the pine nuts they can, well beyond what they need to survive until the next potential cone crop one year hence. They feed on those buried seeds through the entire annual cycle. In June and July, when some of the buried seeds germinate, the nutcrackers eat them as well. Stephen Vander Wall and Harry Hutchins observed that "nutcrackers located germinating seeds soon after the seed stems or hull broke through the soil surface." The foraging birds detached the now-visible seeds, shelled them, and ate the endosperm remaining within. Then, using a sprouting seed as a clue, they dug around in the adjacent soil for cached seeds that still lay dormant. Vander Wall and Hutchins suspected these nutcrackers were actively searching for buried seeds. And why wouldn't they? Finding such seeds would allow them to "cheat" by eating another bird's food stores, something they could not do while all the seeds were concealed underground. Vander Wall and Hutchins suggested that the availability of germinating seeds helps young nutcrackers become independent of the adults. Never having cached any seeds, the young birds do not know where to look for any, so the advertisement of sprouted seeds benefits them especially. To find nutritive endosperm, they must remove it from the opened seed coat within a few days of its emergence from the soil or the growing embryo will absorb it. Fortunately, the melting snow pack may continue to expose seeds throughout the summer, so germinating seeds may remain available until the new seed crop matures in early August.

The availability of germinating and dormant seeds may help nutcrackers through the lean year or two following a mast year. If seeds cached in a mast year lay dormant for two, three, or four years, they would be fed upon throughout the next year. However, they might not be available even to the bird that cached them much beyond that year, because the nutcracker might not remember where it cached them. Balda and Kamil found that nutcracker memory lasts at least 394 days, but at a diminished level. They did not establish its outer limit. However, if memory endured until the second June (about twenty months since seed harvest) a nutcracker could live on dormant seeds and the germinants that advertise them until the next crop is ready. Such a memory, combined with still-dormant seeds stored in the soil, could spell survival for nutcrackers. If this scenario bears any resemblance to reality, whitebark pine's seed dormancy may be an adaptation that benefits the pine in the long run by helping maintain viable populations of nutcrackers.

CHAPTER 13

Origins

EAR THE HAMLET OF EL CHICO, IN THE MEXICAN state of Hidalgo, a range of heavily wooded hills is crowned by the rocky peak of La Muela—the Molar. If you let yourself down a steep slope, slick with recent rains and densely grown with towering *oyamel* firs, you will reach in the *barranca* below La Muela a tumbling clearwater brook. Here, at an elevation of 8,600 feet, the boulder-strewn banks are dominated by groves of white pines with long, slender, blue-green needles, evenly spaced tiers of horizontal limbs, and trunks clothed in furrowed gray bark. Clusters of cylindrical cones, emerald-green in their final summer, hang from branch tips, beads of aromatic resin congealed on their scales. Trunks of the larger pines and firs are encrusted with blue-gray lichens and studded with corsages of red-flowered bromeliads. Their massive lower limbs bear thickly layered sphagnum in which ferns root, forming von Humboldt's forest piled upon forest. The pines are Mexican white pines, *Pinus ayacahuite*, and local farmers call this place "La Ayacahuite." Take away the epiphytes, especially the bromeliads, and this could be an Adirondack haunt or a canyon bottom in northern Idaho. Places like it can be found in several other central and southern Mexican states, and in Guatemala, El Salvador, and Honduras, where the elevation is high, the soils deep and moist, the air cool.

To outward appearances, *ayacahuite* would seem the typical wind pine. Its limbs spread widely, so when the cones open in the fall, their small, long-winged seeds slip freely into the unobstructed breeze that rustles the treetops.

Pinus ayacahuite is the southern anchor of a three-species complex of closely related pines that extend from about 15° to 53° north latitude, an astonishing reach of over 3,700 miles. By traveling northward through this complex, we can substitute space for time, and follow the track of an evolutionary process that still continues.

Two thousand five hundred miles north of La Ayacahuite, in the Canadian province of Alberta, the Bighorn River flows east off the backbone of the Rocky Mountains. On steep granitic slopes rising from its north bank, limber pines (*Pinus flexilis*) grow in scattered, open woodland with white spruce, quaking aspen, and Rocky Mountain junipers. Their multi-forked vertical crown shoots hold aloft for corvid examination circlets of green cones dripping sparkling beads of resin. The cones contain large seeds that are usually wingless, but sometimes have short, blunt wings (Figure 13.1). Limber pine is close to its northern extremity here, but vigorous saplings abound nevertheless, and two-hundred-year-old trees still grow at a fair pace. Seedlings are predominantly stem-clumped in the characteristic way of nutcracker-dispersed pines everywhere, grouped in twos, threes, and fours among the dry grass and fireweeds.

From typical wind-dispersed pine in the south to corvid-dependent pine in the north, this cordilleran series of five-needled pines shows a pattern of

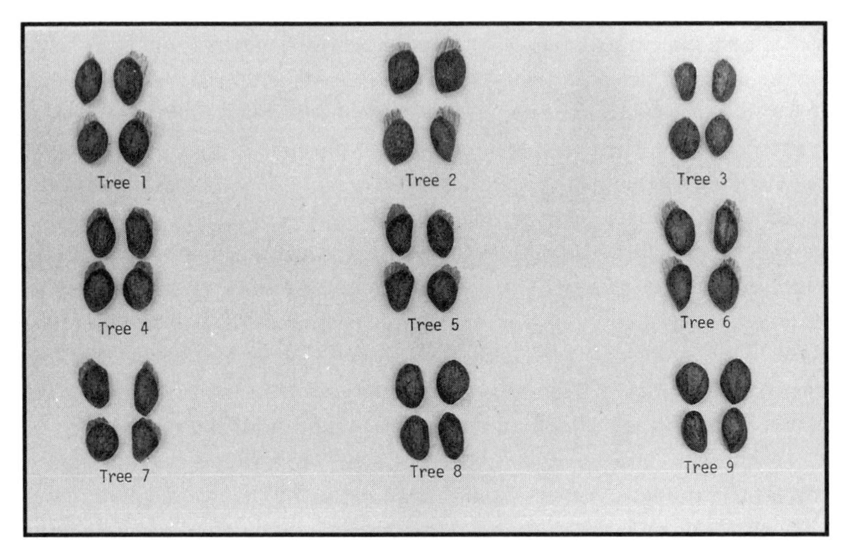

FIG. 13.1 Limber pine seeds sometimes betray their ancestral winged state. A population in the Custer National Forest, Montana had many trees with vestigial seed-wings that are useless for wind dispersal.

continuous variation that signals a close genetic relatedness. In my view, the Mexican white pine is ancestral to the corvid-dispersed southwestern white pine (*Pinus strobiformis*) that lies to its north, and to limber pine, which extends north beyond *strobiformis*. I argue here that *ayacahuite* represents an ancient line and limber pine is its most recent offspring. It wasn't easy for *ayacahuite* or its predecessor to give birth to limber pine, so two midwives helped the process along. One of them was a jay; the other a nutcracker. How could bird pines have descended from a wind pine? By what train of events could new pines arise from an old one? And what is the nature of the evidence that favors such a view?

Assume an ancestral *ayacahuite*, in all respects a conventional white pine—with horizontal branching, pendent long-stalked cones, and small long-winged seeds—growing in moist Mexican and Central American environments. Its seeds are fed upon by a corvid ancestral to the Steller's Jay. Some seeds are cached by the jays, but most are dispersed by wind. Since moisture is plentiful here, cached seeds have no competitive advantage over windborne seeds. Nor do the occasional mutant seeds that are larger or shorter-winged have any advantage over the far more numerous small seeds with long wings. Thus the pine continues to evolve as a wind-dispersed tree of moist sites.

But at the northern edge of its range, the pine encounters drier conditions. Here the cached seeds produce a larger proportion of the seedlings than they did in the moister area, for two reasons. First, because they are sheltered from the drying wind and better able to survive desiccation, and second, because a layer of soil protects them from rodents foraging on the surface. Seeds have unequal chances of being cached, and the likelihood of being cached now becomes an issue of life or death. Jays prefer large seeds to small ones, so they cache large seeds when they can get them. Seeds with short wings, or with none at all, are more likely to remain in open cones long enough to be found by the foraging jays. So over time, caches tend to be made up disproportionately of larger and shorter-winged or wingless seeds. Larger seeds contain more food reserves, so the seedlings that emerge from them are better equipped to survive on the stressful dry sites. Seed size and wing length are heritable characters, so the selective behavior of the jays results in an increase in trees bearing larger seeds with shorter wings.

Typical wind pines carry their branches mainly in horizontal layers, and the cones hanging from their tips are not easily seen by birds flying overhead. But occasional mutants with cones borne on steeply ascendent limbs display their cones effectively to winged foragers, and are likely to have more of their seeds harvested by those foragers. Long cone stalks typify wind pines like the Mexican white pine. These allow the cones to dangle, which makes it difficult for birds to harvest the seeds. Stalkless, or "sessile" cones, on the other hand, are rigidly attached to their branch, and it is easier to remove their seeds. Thus corvid seed harvest, and the subsequent establishment, survival, and reproduction of trees arising from seed caches, drives the morphology of the tree population in the direction of ascending limbs with sessile cones bearing large, wingless seeds.

Raising its cones out from the shade of overhanging limbs into the sun creates a problem for the evolving bird pine. Cones can overheat, denaturing the proteins in their developing seeds. But the evolution of thicker cone scale apophyses with a high moisture content while green, diffuses heat and roughens the cone surface, thus cooling the cones and the seeds within.

Eventually these pines, which have long been in symbiosis with the jays, are encountered by nutcrackers that have arrived relatively recently on this continent and have extended their range southwards. The nutcrackers, always alert to a new food source, now harvest and cache the nutritious Mexican white pine seeds, gradually expanding the pine's range to the north, where open habitats are available on high, dry mountains with cold winters. Repeated caching of seeds on these sites now results in selection for increased cold-tolerance in the continually evolving pine, and adapts it to areas where the growing season is much shorter than in its original southern home.

The nutcrackers exert far more intense selection pressure on the pines than did the jays. Nutcrackers travel in flocks, and they have an obsession for pine seeds. Thus they harvest and cache a much larger proportion of the seed crop. This causes many more seedlings to appear. Every seedling is a unique genotype, and the more genotypes, the more potential variability from which to further select. As the nutcrackers make their choices, they unknowingly mold the germ plasm of the pines they are scattering along the Rockies and across the ranges of the Great Basin. The result is a chain of varying populations that have confused taxonomists and foresters for a century.

All interested parties agree that Mexican white pine and limber pine are "good species," but the legitimacy of southwestern white pine, which lies between them, has long been questioned. Since 1848 it has been regarded first as a good species (*Pinus strobiformis*), then as variety *reflexa* of limber pine, then as *Pinus reflexa*, as variety *strobiformis* of Mexican white pine, variety *reflexa* of Mexican white pine, and as variety *brachyptera* of Mexican white pine! A nomenclatural problem of such long standing usually indicates a confused biological situation. While the southern populations of Mexican white pine and of limber pine to the north are relatively stable and easily recognizable, those from (roughly) Mexico City north into Arizona and New Mexico can be perplexing. Typical *ayacahuite* is followed by a larger-coned and shorter seed-winged form called variety *veitchii*. Beyond latitude 20°

north, this is followed all the way to the U.S. border by a sometimes very large-coned, but nearly wingless-seeded form called variety *brachyptera* by some, *Pinus strobiformis* by others. Jesse Perry, who has probably spent more time with Mexican pines than any other botanist, believes this last stretch of territory contains overlapping populations—even hybrid swarms—of both *strobiformis* and *ayachahuite* variety *brachyptera* (Perry 1991). As I see it, this, the world's greatest north-south chain of pine populations, may have locally bewildering combinations of characters. But taken as a whole, there is a clear trend from wind pine characters in the moist subtropical south to bird pine characters in the cold semi-arid north (Figure 13.2). Precisely where along the chain one draws the lines that separate species and varieties on the map cannot change that salient fact of biology. And for this we have corvids to thank.

There is as yet no definitive evidence for this view of evolution in the *ayacahuite-strobiformis-flexilis* complex. There are, however, several strands of indirect evidence that are consistent with it. For example, it is now believed that during the late Eocene and the Oligocene Epochs (about twenty-six to fifty-four million years ago) major episodes of volcanism and mountain-building in Mexico and Central America created an enormously diverse range of climatic, topographic, and edaphic (soil) conditions there. At that time, ancestors of southwestern white pine and Mexican white pine are believed to have been in southern refugia, escaping the hot, humid conditions further north. When the Eocene ended, the climate cooled rapidly. During the Oligocene, and the Miocene and Pliocene epochs that followed, many new pine species evolved in the diverse Mexican habitats, including new white pines and pinyon pines. It is during the Miocene that the New World jays are believed to have evolved, and that the pinyon pines marched north into northern Mexico and the western United States.

Further, the merging of the three pines into each other, both geographically and morphologically, hardly seems the chance occurrence of unrelated but coincidentally similar species. Far more likely is the hybridizing of species that still retain the ability to crossbreed because their genetic relationship is a close one. Experiments artificially crossing limber pine with southwestern white pine have yielded hybrid offspring. Such "crossability" is regarded by geneticists as evidence of close relatedness.

Another supporting line of evidence lies in the heritability of the characters that distinguish the wind pines and bird pines of this complex. Branch angle is known to be heritable among pines, and conventional white pines worldwide all have upswept limbs in the upper crown. In many of the bird pines, this trait is merely more exaggerated, often encompassing the entire crown. Seed wing characters and seed size are maternally inherited characters.

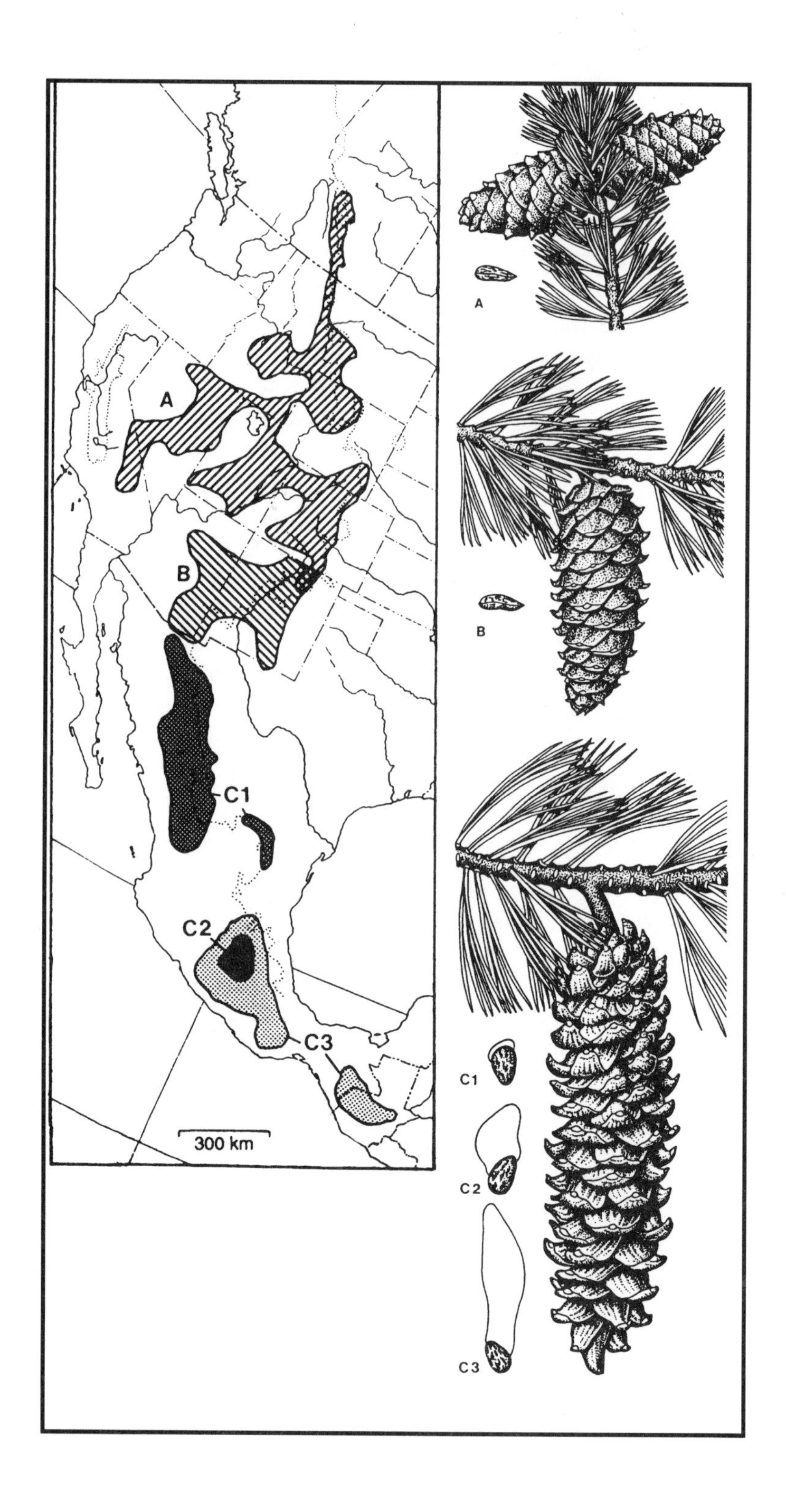

A
B
C1
C2
C3
300 km
A
B
C1
C2
C3

Finally, the availability and capability of corvids that are known to harvest seeds of these pines provides the mechanism for simultaneous selection and dispersal. I have observed Steller's Jays harvesting *ayacahuite* seeds at Laguna de Zempoala in Morelos and Desierto de los Leones near Mexico City. A member of the *Guardia Forestal* at Desierto complained loudly to me about his old adversary, the dark jay with the black *capete*, and fervently wished it would limit its depredations to nearby cornfields. Steller's Jays are present throughout the ranges of all three pines of the complex. Numerous studies have documented harvest and caching of limber pine seeds by Clark's Nutcracker, and research by Craig Benkman and his colleagues (1984) has documented the nutcracker's use of southwestern white pines as well. Thus the essentials of this hypothesis seem to be in place, awaiting confirmation, preferably by molecular genetic studies.

Something similar seems to be happening in Japan, where another bewildering species-complex has long confused foresters and botanists. Japan is a land of rich forests and many picturesque conifers. One of the most beautiful of these is *Pinus parviflora*, the *goyomatsu* or Japanese white pine, a graceful tree distributed widely from the south tip of Kyushu to Hokkaido. This pine is the northernmost of a group making up the "*parviflora*-complex" which also includes three wind pines scattered along the east edge of the Asian mainland from Vietnam to Taiwan. The widespread Japanese white pines have been grouped under the name *Pinus parviflora* since the German naturalists von Siebold and Zuccarini described the species in 1842, but clear differences have long been noticed between its northern and southern populations. Some observers have called the trees north of Tokyo *Pinus pentaphylla*, and those to the south *Pinus himekomatsu*. Others have considered them varieties of *parviflora*. Both varieties, or species if you prefer, have large seeds that one would expect to attract corvids. These seeds are retained in the cone by wings that stick to their cone scale. Compared to their relatives further south, both of the Japanese varieties look like bird pines. And, indeed, Mitsuhiro Hayashida has documented Japanese nutcrackers dispersing *pentaphylla* seeds on the rocky slopes and ridges of Mount Apoi, in Hokkaido. The details noted by Hayashida are reminiscent of others pertaining to nutcracker dispersal elsewhere: seeds were

(*facing*) **FIG. 13.2** The "limber pine complex," consisting of limber pine (A), southwestern white pine (B), and Mexican white pine (C). Southernmost populations are typical Wind Pines in seed and cone morphology and cone orientation. Northern populations are typical Bird Pines. C1, C2, and C3 represent Mexican white pine varieties *ayacahuite*, *veitchii*, and *brachyptera*. By Marilyn Hoff Stewart, with permission of University of Chicago Press.

cached in open areas, caches had one to eleven seeds (average 6.6) and were made two to three centimeters deep, and 37 percent of individual trees tallied along mountain trails were members of stem clumps. Clearly, *pentaphylla* is a bird pine, and *himekomatsu*, whose seed-wings are much the shorter of the two varieties, is almost certainly a bird pine as well (see Figure 2.2). Their close relationship to the wind pines further south suggests they have recently arisen from those species via selection imposed by nutcrackers, in the same way I have speculated for limber and southwestern white pines.

The analogy here with the Mexican white pine—southwestern white pine—limber pine group in North America is further supported by the taxonomic confusion that both groups share. Such disarray, bordering on botanical chaos, can be expected where evolution is progressing rapidly, creating a nursery for new species, and allowing a broad spectrum of variable populations to coexist.

If these pines of the New and Old Worlds are works in progress, the pinyons and stone pines can be regarded as fully evolved bird pines. As pointed out in chapter 2, the pinyons form an ecologically and geographically coherent group of completely wingless-seeded species ranging from central Mexico to southern Idaho. All have large seeds that are held in deep cone-scale cavities by membranous tissue of the scale surface. Frequent hybridization among the northern species of the group, and taxonomic disagreement over how to interpret their variability, indicate that evolution of the group continues actively. The pinyons are usually considered to have spread northward from a Mexican center of origin, and it is in central Mexico where pinyon species diversity is at a maximum. All of Mexico is jay country, and two known pine-nut cachers—Steller's and the Scrub Jay—today range across all Mexican states where pinyons grow. The pinyons are believed to have originated in the late Miocene, and a Pinyon Jay-like corvid left fossil remains in Colorado at about that time. This supports a hypothesis that pinyons were created by the selective action of jays, because it shows that a potential pine-nut-harvesting jay was available at about the right time in about the right place.

The stone pines of subsection *Cembrae* form another group of bird pines that share a syndrome of specialized traits that facilitate the seed harvesting activities of their avian seed dispersers. All of the stone pines have wingless seeds retained within non-opening cones. The scale tips of all species are easily broken off by a nutcracker's bill, exposing the seeds, which are held in the cone core until removed. The cones are almost entirely borne on the tips of ascending limbs, so they are clearly visible to flying nutcrackers, or birds perched on the crown; and the branch tips form a stable platform from which to forage. The cone-bearing surface may be a meter off the ground in Japanese stone pine, or thirty meters in old-growth Siberian or Korean stone pines.

Limber pine's crown form is virtually identical to that of whitebark pine, and whitebark's is the most radically forked and verticalized of all the stone pines. Partly for this reason some researchers have felt that whitebark and limber pines must be closely related, that perhaps whitebark was the ultimate product of the *ayacahuite* line. But chloroplast DNA and isozyme studies by Konstantin Krutovskii and his colleagues have shown that whitebark pine clearly belongs with the stone pines, and limber pine with subsection *Strobus*. Thus the close resemblance of these western American pines is the result of convergent evolution, not common ancestry. The stone pines may stem from an ancestral version of the Himalayan blue pine, or from the Balkan white pine, or from a species that has left no wind pines at all in its legacy. The origin of the *Cembrae* pines has been little investigated and remains shrouded in mystery.

But regardless of where the stone pines originated, it is apparent that North America is where they finally ended up. Why should one assume the *Cembrae* originated in Eurasia and then entered North America, presumably across a Beringian land bridge, and not vice versa? I would argue this on two grounds: one arboreal, the other avian.

The profusion of stone pines in Eurasia (four species with at least six named subspecies, nine cultivars, and several botanical forms and varieties) suggests a long stone pine presence on that land mass. In contrast, whitebark pine is the only stone pine in North America; it has no named varieties or subspecies; and only one cultivar has been selected. This relative uniformity suggests a recent arrival.

Krutovskii, Politov, and Altukhov further support the hypothesis that whitebark pine is a recent cross-Beringian immigrant from northeast Asia. Their molecular genetic data indicate that whitebark pine diverged from the Eurasian stone pines between 600,000 and 1.3 million years ago. The Bering Strait is estimated to have opened 1.8 to 3.5 million years ago, putting an end to further migration across the land bridge. It is therefore reasonable to guess that a pine very similar to today's Siberian stone pine was brought into Alaska while the Bering Strait was still closed, and then became isolated from Asia when the strait opened. It subsequently differentiated into what we now recognize as whitebark pine.

The avian evidence parallels that of the trees. According to the Russian ornithologist Dement'ev and his colleagues, the Eurasian Nutcracker comprises ten subspecies resident from Scandinavia to Kamchatka, from the Arctic Circle to the Tropic of Cancer (Figure 13.3). Here again, the great

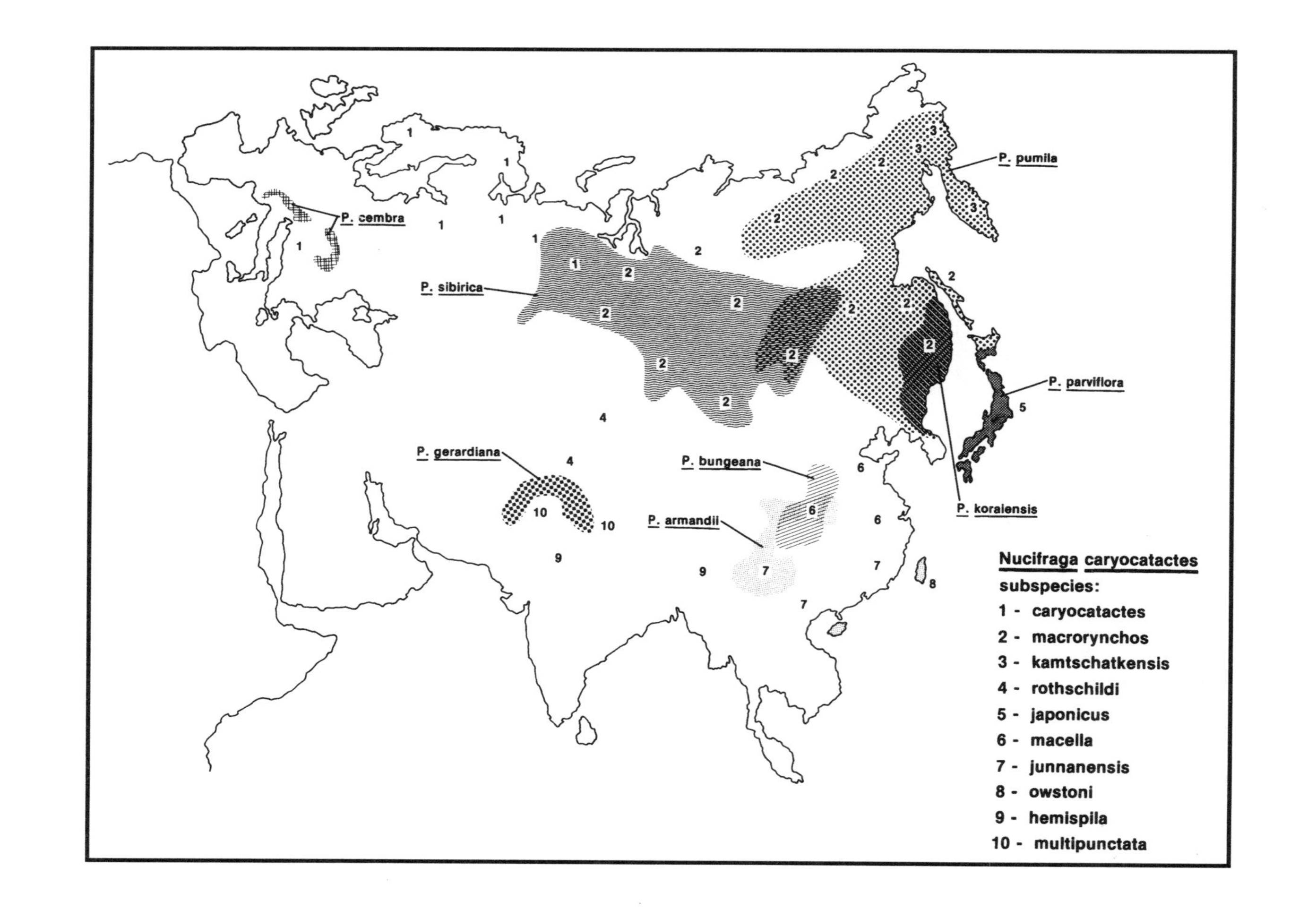
P. cembra
P. sibirica
P. pumila
P. parviflora
P. koraiensis
P. gerardiana
P. bungeana
P. armandii
Nucifraga caryocatactes
subspecies:
1 - caryocatactes
2 - macrorynchos
3 - kamtschatkensis
4 - rothschildi
5 - japonicus
6 - macella
7 - junnanensis
8 - owstoni
9 - hemispila
10 - multipunctata

variability of the Eurasian species contrasts with Clark's Nutcracker's lack of any named subspecies or variety in North America. The differentiation and subsequent spread of the Eurasian Nutcracker into various forest types suggests a long period of occupation on that giant land mass, while the uniformity of Clark's Nutcracker indicates it is a relatively recent arrival in North America. The unprovable conclusion suggested by these parallel stories is that the pine was brought to North America by the nutcracker, and that both then differentiated new species.

The remaining wingless-seeded soft pines—chilgoza, lacebark, and Armand pine—are all found within the range of the Eurasian Nutcracker. Of those Asian trees, only Armand pine has no morphological adaptations to help keep the large, heavy seed in its cone. In this regard it is similar to limber pine, which, despite its dependence on Clark's Nutcracker, has nothing to show for that dependence in the structure of its cone.

These facts, and others presented earlier, strongly support the conclusion that the wingless-seeded soft pines—about twenty-two species in the Old and New Worlds—have evolved through the selective action of corvids.

In a sense they have been domesticated by jays and nutcrackers from wild ancestral stocks that were far less valuable as food sources. And this has happened not once, but many times during the past several million years. Each type of cone adaptation has probably resulted from a separate evolutionary event. Thus, the pinyon pines have a common ancestor in which seed-retaining cone-scale tissue evolved, and the stone pines have a common ancestor in which non-opening cones of non-fibrous scales evolved. Not enough is known about chilgoza and lacebark pines to say with confidence whether their different cone adaptations (see chapter 2) rebut the logic of placing these species in the same subsection (*Gerardianae*), implying that they have a common ancestor. The seed wing of Japanese white pine behaves like that of chilgoza pine. Whether that is true of both the *pentaphylla* and *himekomatsu* populations, and whether it signals a common lineage of these very different pines, cannot now be said.

One final evolutionary deduction about pines that can be drawn from these relationships deals with the relative "primitiveness" of the various cone and seed morphologies in the subgenus *Strobus*. George Russell Shaw, the most influential of twentieth-century pine specialists, considered the wingless seed and non-opening cone of the *Cembrae* pines to be the primitive

(*facing*) **FIG. 13.3** Distribution areas of the ten Eurasian Nutcracker subspecies (Demet'ev et. al. 1954) and wingless-seeded Eurasian pines (Critchfield and Little 1966).

condition within the genus. According to this view, the most evolved characters in *Strobus* are the effectively winged seeds and opening cones of the subsections *Strobi* and *Balfourianae*. But if corvids are responsible for the characters Shaw called primitive, those characters can only be as old as the corvids. The oldest corvid fossil that closely resembles present-day seed cachers is the Upper Miocene *Miocitta galbreathi* from Colorado, a bird similar to the nutcrackers and pinyon jays. The age and location of that fossil suggest that pinyon pines and limber pine may not be much over twenty million years of age, hardly old enough to qualify as primitive in a genus that was already established 130 million years ago.

I find it more logical to regard winglessness as an advanced specialization than as a primitive character for three reasons:

- Only pines have wingless-seeded species. All the conifers that diverged from the pines have only winged seeds. If winglessness were primitive, we would not expect it to be absent from all of those genera.
- It is difficult to imagine how an early wingless seeded species, especially one with a *Cembrae*-type indehiscent cone, could have survived for over one hundred million years before the advent of highly specialized corvids. How would it have been dispersed? Why would there be no fossil record of such seeds or cones earlier than the Miocene?
- The data on corvid-pine interactions provide a consistent and logical model for the evolution of specialized corvid-dispersed pines from more generalized wind-dispersed types by basic natural selection mechanisms.

The mutualism of corvids and bird pines tells us there is still plenty of inherent vitality in this ancient line of conifers. Many botanists have written the conifers off as a fading tribe of primitives undergoing replacement by the more modern, therefore presumably more survivable, flowering plants. But the relatively recent evolution of numerous bird pines shows that *Pinus* is still vigorous enough to be opportunistic, and can find new ways to solve old problems.

And what of the pine birds? Surprisingly, far less research has explored the origins and evolution of corvids than of pines. In *The Pinyon Jay*, Marzluff and Balda reviewed evidence bearing on jay and nutcracker evolution. The following discussion is a summary of some of their main conclusions. About thirty-five million years ago the New World jays evolved on the land masses destined to become the Americas. Meanwhile, on what would later become Eurasia, the Old World corvines, consisting of other jays, magpies, crows, ravens, and nutcrackers, evolved in isolation from the American jays. During the Pleistocene ice age, when a falling sea level exposed the Bering land bridge linking Asia and North America, Gray Jays, ravens, crows, magpies, and nutcrackers found their way into the New World. In the past, bird taxonomists have been baffled by the physical resemblance of the Pinyon Jay to Clark's Nutcracker, whose roots lie in the Old World. This led them to believe that Pinyon Jays were descendants of Old World jays. But in common with those jay species acknowledged to be of American origin, the nestlings of Pinyon Jays do not have a downy covering like that of nutcrackers, crows, and ravens. The Pinyon Jays also share with other New World jays an adaptation of the jaw articulation that aids in opening acorns, a character not found in Old World corvids. Thus it appears that the physical similarity of Pinyon Jays and nutcrackers is the result of convergent evolution, not of common descent.

The origin of the two nutcracker species has been mildly controversial. The Soviet ornithologist Stegmann (1934), refuting a colleague, argued that Clark's Nutcracker is more primitive than the Eurasian species, basing his conclusion entirely on bill characters. The American ornithologist Dean Amadon (1944), disagreed, on the grounds that the nutcracker is a specialized offshoot of the Old World jays, and that it "reached North America long enough ago for the American species to become very distinct from the Palaearctic one." Amadon's version is consistent with the most likely stone pine scenario. There is a rough correspondence between wingless-seeded pine species in Eurasia, and the subspecies of its nutcracker disperser (Figure 13.3), but no evolutionary conclusions have yet been drawn.

Vander Wall and Balda have proposed a model to show how corvids may have evolved from relatively unspecialized seed-cachers to more specialized species with greater dependence on cached seeds. Because greater dependence on the cached seed resource makes the corvid more vulnerable to a failure of the cone crop, it acts as a positive feedback increasing the efficient use and recovery of cached seeds. This feedback response would place a premium on the evolution of morphologies and behaviors that make the corvids better pine birds. Opposed to this tendency is the negative feedback resulting from increased vulnerability to seed shortages. This puts a premium on the adaptive response to food shortage that characterizes Pinyon Jays and nutcrackers—long eruptive flights in search of alternative foods. Vander Wall and Balda suggest it is the point of balance of these two feedbacks that determines how dependent on its stored food a corvid can afford to become. And this level of dependency is a major factor in the corvid's evolution of behavioral adaptations. This model can be viewed as the "bird half" of a coevolutionary model of pine birds and bird pines.

Coevolution refers to the postulated mutual selective effects on one another of two interacting populations. To assert that a corvid species and a pine species are coevolved is to say that one of those species brought about changes in the other by selection, and was then itself selected by the very changes it elicited. Stone pines harbor a syndrome of traits that raise the efficiency of nutcracker foraging. These include verticalized fruiting branches, sessile cones, non-opening "breakaway" scales, seed-retaining cone cores, and large seeds that are wingless as well. These characters increase the foraging efficiency, and thus the fitness, of nutcrackers, including those bearing such traits as longer, stronger bills, larger sublingual pouches, and better-developed spatial memory. Since those nutcracker traits make the seed-harvesting and caching enterprise more reliable in ensuring pine regeneration, they not only enhance the survivability of the nutcrackers that bear them, but that of the pine as well. If nutcrackers preferentially harvest the seeds of those individual pines that most reward them, they increase the frequency of the facilitating pine traits; thus they become agents of natural selection and channel the pine's evolution. Meanwhile, the pine is doing the same thing to the nutcracker. Coevolution is probably at work in all mutualisms that have a long history, including not only those of nutcrackers and stone pines, but probably Pinyon Jays and pinyon pines as well. Looser mutualistic interactions like that of nutcrackers with pinyon pines and limber pine are probably not coevolved, but one-sided. The nutcracker, after all, was already a pine bird when it discovered its first pinyon and limber pines.

Whether a mutualism is coevolved or not is probably not subject to proof. The question may occasionally stimulate the flow of blood through the veins of argumentative scientists, but influences not one whit the fate of the mutualists themselves. What does determine the fate of a mutualism is the health and well-being of each of its members. When one member of a mutualism catches cold, the other sneezes. Some serious sneezing may soon be heard in North America, because a critically important mutualist has caught a cold.

CHAPTER 14

Is the Keystone Slipping?

ONSERVATION BIOLOGISTS HAVE IN RECENT years adopted the concept of the "keystone mutualist" to call attention to interdependence among the plants and animals of an ecosystem. A keystone mutualist is so closely involved with other organisms that if it becomes extinct, or even seriously depleted, the effects will ramify throughout the ecosystem. Norman Myers has probably expressed this concept most starkly:

> If a keystone mutualist is eliminated as a result of human disturbance of forest ecosystems, the extinction of several other species will follow inevitably. Still more to the point, these additional losses may, in certain circumstances, trigger a cascade of linked extinctions. Eventually a series of forest food webs can become unravelled, with shatter effects throughout their ecosystems. (Myers 1986, p. 403)

While it may be difficult to identify many plant species as keystone mutualists, surely a likely candidate for that status is whitebark pine. Its importance as a food source for nutcrackers, squirrels, and bears, and its role in forest succession argue powerfully that whitebark pine is not just another tree species, but that it occupies a central biological position in relation to many of its neighbor organisms. If whitebark pine were to disappear over large areas where it is currently distributed, there would be no pine nuts there. Nutcracker populations and Red Squirrels would be displaced. There would be no rich fat source for hibernating bears. Even before the pine disappeared from an area, it would probably reach a density too low to attract the nutcrackers needed to disperse its seeds and establish its seedlings. Later, when only isolated trees still survived, most of their seeds would result from self-pollination, reducing both the numbers of seed and the competitive ability of their seedlings. Thus the local loss of whitebark pine sets in motion a positive feedback that makes the pine increasingly unable to become reestablished, resulting in eventual "local extinction" of the pine, its disperser, and flora and fauna dependent upon it. Unfortunately there is reason to fear that such a doomsday scenario has already been set in motion for whitebark pine, that this keystone of many of our subalpine forests may indeed be slipping.

Whitebark pine faces four threats: competition from more shade-tolerant trees due to fire exclusion; heightened bark beetle attacks, also engendered by fire exclusion; loss of habitat through global warming; and quick death from a parasitic fungus. All these threats are the inadvertent effects of human action. All could, conceivably, be lifted by human action. But the will and the commitment to do so would have to be stronger than anything we have seen in the past when nature has been threatened, and time is running short.

The price of fire exclusion and control in our forests has been high. Throughout the mountains of western North America, the efficient twentieth-century suppression of fires has resulted in the abnormal survival of many millions of shade-tolerating trees, mainly firs, beneath canopies of early-successional pines. In the past, when Europeans were still an ocean away and lightning-strike fires ran free, thin-barked understory trees died young in the numerous ground fires that recurred at short intervals, keeping the forest floor relatively clear of burnable debris. In the West the forest was almost park-like throughout the 1800s, and John Muir could write that "the inviting openness of the Sierra woods is one of their most distinguishing characteristics." That was when ponderosa pine forests burned every fifteen years on average in the Northwest and every four to ten years in the Southwest, when whitebark pine in the northern Rockies often had a fire-return interval of thirty to a hundred years, and lodgepole pine sixty to eighty years. Under such regimes firs were short-lived, and fires tended to burn close to the ground in the organic litter accumulated since the previous fire. In today's forests there is more accumulated fuel allowing a hotter burn, and understory firs often become fire ladders, conducting the flames upward into the forest canopy to produce explosive crown fires. Fire control in recent decades has allowed firs and spruces to survive in greater numbers, and to create deeply shaded conditions generally unfavorable for whitebark pine growth. In addition, it has allowed highly incendiary conditions to build up in lodgepole pine (*Pinus contorta*) forests just down-mountain from whitebark pine stands, abetting the upward spread of destructive fires. Further, the old lodgepole forests now protected from fire are subject to severe outbreaks of mountain pine beetle, a bark beetle that then spreads upward into the whitebark pine zone.

One of many such outbreaks occurred in Montana's Whitefish Range during a period lasting from 1979 to 1985. A beetle infestation that began in lodgepole forests of Glacier National Park killed whitebark pine in a 25,000-acre area of the Park and adjacent Flathead National Forest. Ironically, the suppression of low-intensity fires has increased whitebark pine's risk from truly destructive fires, and from mountain pine beetles (*Dendroctonus ponderosae*) as well. There is little reason to expect this situation to improve in the short run. As the extensive lodgepole pine stands that became established in the northern Rockies after the great fires of 1910 reach maturity, further beetle outbreaks can be expected to spread into the whitebark pine zone.

The potential effects of global warming on whitebark pine are speculative, depending on the responses not only of whitebark pine, but of Clark's Nutcracker as well. Additionally, the behavior of associated tree species, and the changed insect, disease, and fire situation will also have a powerful impact on whitebark pine and all other inhabitants of subalpine ecosystems. In general, whitebark pine will experience warmer and drier conditions at its lower elevations. It may not be able to tolerate such droughty environments, and may respond by dying back. Perhaps it will survive, but at a lower level of vigor. It will find itself under increasing pressure from fire and mountain pine beetles as the lodgepole forest migrates upslope. If Clark's Nutcracker cooperates by caching seeds at ever-higher elevations in pace with the warming climate, whitebark pine's elevational range will gradually be displaced upwards. After the Glacial Maximum nutcrackers apparently "helped" singleleaf pinyon to migrate northward across the Great Basin, so there is precedent for such movements. But then the nutcracker had several thousand years in which to adjust to climatic change, while the changes that lie ahead are expected to be much more rapid. Singleleaf pinyon migrated across a land surface comprising tens of thousands of square miles: there was always lots of potential pinyon habitat further north as the glacial meltdown allowed the climate to warm up. But whitebark pine, already isolated high in the mountains, has nowhere to go but up. Where it is already at the summits of ridges and mountain peaks, it cannot expand upward, and can only remain static above while being decimated below. Where the mountains rise far above the present zone of whitebark pine, as in the Cascades, Sierra Nevada, and Canadian Rockies, much of the available upslope habitat will be on steep faces of bare rock. Even with the best of nutcrackers' intentions, such sites cannot be expected to support biologically meaningful forests of whitebark pine. So, if the currently-held views of global warming are borne out, it is difficult to visualize the whitebark pine ecosystem as a viable component of the environment. Whitebark pine's long-term prospects do not look promising.

Unfortunately, whitebark's long-term prospects may not even be relevant. The species may not survive into the long term if short-term catastrophe overtakes it. The stevedores who off-loaded a thousand pine seedlings from a Romanian freighter to the docks of Vancouver, British Columbia in 1910 could not have known they were setting the stage for a biological disaster in the coniferous forests of western North America. The crated seedlings were eastern white pines shipped from a French nursery and infected with that pox of the five-needled pines, white pine blister rust. The causative fungus, *Cronartium ribicola*, produces spores on the leaves of *Ribes* shrubs—gooseberries and currants, wild or domesticated—which are brought to the needles of nearby susceptible soft pines by late-summer winds. If the weather is suitably cool and humid, spores will germinate and penetrate the pine's stomates, producing a filamentous hypha that grows, wormlike, down through the needle, then enters the woody twig and branch system where it causes swollen, discolored areas to form on the bark. In subsequent springs, raised blisters push through the bark to discharge aeciospores, which when windborne can infect *Ribes* plants hundreds of miles away. As the blisters deflate, patches of bark die, and the fungus spreads into adjacent healthy bark, enlarging the canker. Infected branches die, producing the characteristic red "flagging" of dead limbs scattered in the tree's crown. Infected trees die when the trunk becomes encircled by cankers.

Within fifty years white pine blister rust had spread from its point of entry at Vancouver throughout the range of western white pine in British Columbia, Washington, Oregon, Idaho, and Montana. In northern California sugar pines were quickly succumbing to the invading fungus, though for a time it seemed that chronic summer drought south of the 38th parallel would protect trees of the southern Sierra Nevada. It did not. The rust has moved explosively even into old-growth sugar pines in that area. Blister rust appears on limber pine east of the Continental Divide in Montana, and has become widespread in whitebark pine, where by 1959 it was reported to have compromised the watershed-protection value of this tree in parts of three northwestern states.

Cronartium ribicola's stunning success as a killer of North American white pines was aided by the variety and abundance of available hosts, both white pines and the *Ribes* growing among them, and a climate suitable for spore germination over large areas of mountain country. Natural resistance to the disease is rare among the North American white pines. These trees have evolved in the absence of *Cronartium ribicola*, and thus have not been subject to natural selection for resistance to the pathogen. That is why North American white pines are much more susceptible than the Eurasian species that have coevolved with the fungus. Here, for example, are relative resistance rankings of the *Cembrae* pines that have been tested:

Swiss stone pine	1
Korean stone pine	2
Siberian stone pine	4
Whitebark pine	11

Whitebark pine is the species most highly susceptible to white pine blister rust. Among the numerous locales where it has been infected are the northern Cascades of Washington, the Yellowstone area, the Cabinet, Mission, and Bitterroot Mountains of Montana, Crater Lake, Oregon, and even its isolated stands in the Olympic Mountains of Washington. In nine western Montana study areas, the basal area of whitebark pine—a measure of tree size and density—declined an average of 42 percent in the twenty-year period from 1971 to 1991. Where infection was found in only one of seventeen study plots in 1971, by 1991 all the plots had infected trees. At the latter date, an average of 89 percent of the surviving trees were infected by the rust, and crown-kill from the disease averaged 46 percent. Nearly all of the infected survivors are expected to succumb within thirty years. In Glacier National Park, an important Grizzly Bear habitat, Kendall and Arno estimated that 90 percent of the whitebark pines had died by 1990.

Can anything be done to prevent the loss of this keystone mutualist and the organisms dependent upon it? Where successional replacement by firs is underway, land managers can allow light fires to burn, or they can use controlled fires to burn out the firs and favor the pines. Whitebark pine seedlings can be planted to make up for the lost regeneration opportunities of the past several decades. Return to a natural fire regime in downslope lodgepole forests would ensure that most fires there would be light ones, and less of a threat to the whitebarks up-mountain. This would also reduce the likelihood of catastrophic mountain pine beetle outbreaks, but it is a long-term strategy.

Global-scale climate change can not, of course, be realistically addressed at the local level with reference to the needs of individual species. Given a continually expanding world and national population, and sustained production of enormous quantities of greenhouse gases, perhaps our most realistic hope is that whitebark pine harbors unseen genetic adaptability to greenhouse conditions, and can weather the climatic changes. A fallback position is the hope that predictions of the rate of warming will prove oversimplified and fallacious, that climate change will prove a will-o'-the-wisp.

And the white pine blister rust? How can we address this threat? First it should be understood that we have a tripartite complex here consisting of a pine, a *Ribes* alternate host, and the fungus. Past attempts to break up the complex by destroying gooseberry and currant bushes failed disastrously. It simply is not feasible to grub out all the plants of one or several sprouting species over huge tracts of wildlands, no matter how many "blister-busters" are recruited from the campuses and skid rows of America. Nor can control be realized by acting upon the fungus. Conceivably one could genetically engineer a non-virulent strain of *Cronartium ribicola* and hope it would outcompete and replace extant strains, but such a futuristic solution is beyond present-day abilities, especially over vast land areas. Thus the answer must be found in the pines.

Perhaps the most direct approach is to prune infected branches. This has been done with some success in infected stands of western white pine, and might work in whitebark pine, whose cankers grow more slowly. Considering the size of the land area involved, and the ability of the rust to spread explosively, this approach seems analogous to a thumb stuck in the hole in a dike. But, perhaps.

An obvious line of attack is to breed resistant strains of whitebark pine. Forest geneticists have a half-century of experience in resistance breeding, and the United States Forest Service has already begun a modest program based on proven techniques. There is evidence that, despite whitebark pine's high degree of rust susceptibility, about 5 percent of trees in even the most devastated populations have some natural resistance to the pathogen. These can be used as a base population for breeding a more resistant generation, but this is a laborious task that will require considerable research and development to finally become a reality. A variant breeding method would be to create resistant seedlings by hybridizing whitebark pine with Swiss or Siberian stone pine, to capture the exotic species' natural resistance. This too would be a time-consuming, research-intensive process, requiring detailed genetic knowledge of both parent species and their hybrids, as well as the hybrids' reaction to native pests and environmental conditions.

The same result could be sought by using cultural methods like controlled burning and planting to increase the numbers of whitebark pines in the landscape, and then letting nature take its course. The rust would hit, most of the trees would die, but those with some resistance would form the nucleus of a new breeding population. Here nature would do the breeding, in effect recapitulating the selection process that has occurred in Eurasia. But by the time the resistant trees matured, there might not be any Clark's Nutcrackers, Red Squirrels, or Grizzly Bears left to utilize their seeds.

Or we could give up on whitebark pine entirely, and try to replace it with planted Swiss or Siberian stone pines that are already resistant. This switch might even fool nutcrackers, with the result that naturally regenerated forests would replace the fallen whitebarks. Again, a lot of research would be needed to establish feasibility. Even if it was found that a foreign species could fill whitebark pine's biological role, this solution might not be acceptable in national parks, where policy prohibits the use of non-native species in wild areas.

Retired Forest Service geneticist Ray Hoff expresses optimism about the prospects of beating the blister rust by breeding resistant whitebark pines. He has been a key player in the breeding program aimed at producing rust-resistant western white pines at the Intermountain Research Station's laboratory in Moscow, Idaho. Having brought western white pine within forty years from the precipice of economic doom to the release of highly rust-resistant seedlings, Hoff sees the whitebark pine saga as one with a potentially happy ending. Whether that potential is achieved, however, is highly uncertain. On the one hand, the biology has to come out right. Resistant pines must have several resistance mechanisms, to protect them from a notoriously mutable and adaptable rust fungus. The current whitebark pine populations must persist for the century or so it may take for planted resistant trees to produce cone crops sufficient to sustain nutcracker populations. The rust must not continue to scythe its way through vulnerable whitebark pine populations so explosively as to overwhelm all efforts to blunt its force. Meanwhile, the commitment to maintaining this lone American stone pine in its subalpine ecosystem must be firm and consistent. Talent must be mobilized. Long-term support must be assured.

Then, if all goes reasonably well, if no unforeseen disasters nullify our best efforts, perhaps future generations of North Americans will continue to experience the thrill of seeing flocks of nutcrackers raiding whitebark pine seed crops, and the chill of coming upon crushed purple cones lying in freshly turned earth.

NOTES

CHAPTER 2

For a charming account of pines from cultural and scientific perspectives, see *The Story of Pines* by Mirov and Hasbrouck (1976). The most comprehensive modern reference on pines is Mirov's *The Genus Pinus* (1967). The classification of the pines followed here is that of Critchfield and Little (1966), and Little and Critchfield (1969). A third subgenus (*Ducampopinus*) consists only of *Pinus krempfii*, a bizarre rarity endemic to Vietnam. For a survey of other recent classifications of the pines, see Millar and Kinloch (1991).

The reclassification of numerous fossils, especially those in a large nineteenth-century collection, and the discovery of others, has resulted in a much more complete understanding of pine evolution. Millar (1993) has synthesized these data with current knowledge in plate tectonics, radioisotope dating, and paleoclimates. Based on this synthesis she hypothesizes that pines moved into three major refugial areas of the Northern Hemisphere during the Eocene epoch of the Cenozoic era, and that it was in these refuges that much of the diversity evolved that we see in the genus today.

An experiment I performed (Lanner 1985) with long-winged seeds of Himalayan blue pine showed their average rate of descent in still air to be 1.01 meters per second; but when the wings were broken off, the same seeds fell 4.06 meters per second—four times as fast.

Table 2.1 differs from Critchfield and Little (1966), and Little and Critchfield (1969) in the following details: *Pinus longaeva* is added to the *Balfourianae*; the name *P. juarezensis* is substituted for *P. quadrifolia*, which I consider a hybrid of *P. juarezensis* × *P. monophylla* (Lanner 1974); *P. johannis*, *P. discolor*, and *P. remota* are added to the subsection *Cembroides*; and Chiapas white pine is listed as a species, *P. chiapensis*, instead of a variety of *P. strobus*.

Various aspects of pinyon pine biology, history, and cultural involvement are covered in my book *The Piñon Pine: A Natural and Cultural History* (Lanner 1981).

The Italian stone pine (*P. pinea*) is not a member of the stone pine group, notwithstanding its common name. Nor is limber pine (*P. flexilis*), which is erroneously lumped with whitebark pine by Harlow, Harrar, Hardin, and White (1991). The information on stone pines is summarized from Lanner (1990a).

If variety is the spice of life, the pines are clearly the spiciest of gymnosperm genera. The ecological diversity of North American pines has been attributed by McCune (1988) to the opposed forces of constraint due to shared ancestry, and divergence caused by the different selective pressures imposed by diverse environments.

CHAPTER 3

It is unclear who was really the "discoverer" of whitebark pine. According to Sargent (1897), John Jeffrey's collection of specimens was the first. It is from Sargent that I have taken the Jeffrey quotation. Bailey (1975) claimed that microscopic analysis of needles collected by John Charles Frémont in August, 1842 in the Wind River Mountains of Wyoming has "established" that material as whitebark pine, and that Frémont was therefore an earlier discoverer than Jeffrey. The foliage in question had been thought to be from limber pine, which is also found in the Wind River range. But Bailey neither provides nor cites any specific supporting evidence of what was seen in the microscope, by whom, or when. He also suggests that a possible earlier discovery was made in 1826 by Thomas Drummond, who apparently did not collect any specimens, but who mentioned seeing mutilated cones on the "height of land" between British Columbia and Alberta. However, both limber and whitebark pines grow in that area, and nutcrackers routinely "mutilate" unopened limber pine cones to extract the seeds. So if one is to speculate, why not speculate more tidily that whitebark pine and Clark's Nutcracker were co-discovered by Captain Clark in a single *coup d'oeil*? The details of this encounter are described by Cutright (1969).

The fragility of whitebark pine cones and their subsequent disintegration are in sharp contrast to the behavior of those of limber pine, with which whitebark pine is often confused, even by western foresters. Both species are five-needled, both develop broad crowns of forked multiple stems bearing upswept branches, both form relatively open groves on high, windswept sites in the northern Rockies, and both depend on Clark's Nutcracker for seed dispersal. But where whitebark pine has brilliant crimson pollen cones and seed cones that go from purple to brown as they mature, the corresponding organs of limber pine are pale yellow, and green turning to a golden brown. Both species have smooth gray bark, sometimes blushed with pink, on young trees and young branches, but on older trees the rough scaly bark is whitish-gray on whitebark, and brownish-gray on limber pine. The species can be reliably separated by microscopic anatomy of the needles (Harlow 1931), but that is not practical in the field. The use of needle resin canal location has been shown unreliable by Hendrickson and Lotan (1971). Therefore, the best diagnostic feature in the field is the presence or absence of cones and cone fragments. Limber pines of bearing size will always drop large numbers of big, woody, open-scaled cones, which will be found on the forest floor in various stages of weathering, but still in one piece. Whitebark pine cones will almost always be missing, until one gets down and scrapes from the ground the scattered scales and cores broken apart by nutcrackers and squirrels. This was the reason for Newberry's difficulty. Data on variation in whitebark pine cone crops have been documented by Mattson et al. (1994).

The exceptions to the two-year cone development period are Chihuahua pine (*Pinus leiophylla* and its var. *chihuahuana*) of the American southwest and Mexico, and Italian stone pine (*Pinus pinea*) of the Mediterranean area. These unrelated species require about a year longer to produce mature seeds (Mirov 1967).

According to Christensen and Whitham (1991) "The masting habit of pinyons may not solely function to satiate seed predators, but may also have evolved to ensure successful dispersal." During a four-year Arizona study they discovered a most interesting relationship among Colorado pinyon pine, *Dioryctria* cone moths, and the suite of avian seed dispersers—Clark's Nutcracker, and Pinyon, Scrub, and Steller's Jays. The birds foraged most persistently from stands with large cone crops, and from the trees within those stands that bore the most cones. If a stand of trees was heavily attacked by cone moths, and therefore had few good cones to forage from, the birds would avoid even the occasionally cone-laden tree in that stand. Christensen and Whitham thus suggest that the interaction of cone moth abundance and bird behavior acts selectively to favor the trees maturing the largest cone crops, "resulting in a trend toward increased energy investment in fecundity."

Details of pine cone anatomy can be found in Shaw (1914). Harlow, Coté, and Day (1964) demonstrated the cone opening mechanism in pine cones.

The standard source for such details as weights of forest tree seed is the Forest Service seed manual (USDA Forest Service 1974). Data on nutrient content of various species of pine nuts are from Farris (1983), McCarthy and Matthews (1984), Yoon et al. (1989), Lanner (1981), and Lanner and Gilbert (1994). Calorie contents are from Lanner (1982), Hutchins and Lanner (1982), and Vander Wall (1988). In gymnosperms the seed nutritive tissue is technically the female gametophyte.

In several wingless-seeded pines the end of the seed from which the root emerges tapers to a fairly sharp point. This raises the question of whether corvids preferentially cache such seeds root-end down, to the possible, though undemonstrated, benefit of the germinating seedling. According to Jim Todd of Salt Lake City, Utah, who feeds pine nuts to Pinyon Jays in his backyard, most pine nuts are indeed cached in this way.

CHAPTER 4

Some good general references on corvids are Goodwin's *Crows of the World* (1976), Wilmore's *Crows, Jays, Ravens and Their Relatives* (1977), and Angell's *Ravens, Crows, Magpies, and Jays* (1978). Heinrich's *Ravens in Winter* (1989), Kilham's *The American Crow and the Common Raven* (1991), and Marzluff and Balda's *The Pinyon Jay* (1992) mark a recent upsurge of interest in these most interesting birds.

Surprisingly little has been written about the mutualistic relationships that have developed between jays and the Fagaceous trees that produce acorns or beechnuts. A careful behavioral study of the European Jay as a disperser of pedunculate oak, sessile oak, and red oak acorns has been made by Bossema (1979). In the United States, Darley-Hill and Johnson (1981), Johnson and Adkisson (1985, 1986), and Johnson and Webb (1989) have shown the Blue Jay to be highly effective in dispersing oak acorns and beechnuts in the fragmented landscapes of Wisconsin. There are many parallels between these mutualisms of corvids and trees of the beech family, and those described in this book between corvids and pines.

John Grinnell, the famous American naturalist of the early twentieth century, called attention to the role of acorn-cachers like the Scrub Jay in extending plant species' ranges uphill, against the pull of gravity on fallen acorns (Grinnell 1936). In view of what appears to be a limited ability of Scrub Jays to manipulate pine nuts, it is hard to know what to make of Yeaton's (1983) observation that this bird harvests the large, thick-shelled seeds of gray pine.

Marzluff and Balda should be consulted for further details of the behavior and life history of the Pinyon Jay, as should the series of papers by J. David Ligon (1971, 1974, 1978), Ligon and Martin (1974), and Ligon and White (1974).

Balda (1987) has written a useful summary of the impacts of birds, including corvids, on pinyon-juniper woodlands. The comparative study of the three western U. S. jay species and Clark's Nutcracker can be found in Vander Wall and Balda (1981).

CHAPTER 5

In Europe, the nutcracker is known by many names: *casse-noix* (France), *nocciolaia* (Italy), *cascanueces* (Spain), *notenkraker* (Holland), *nötkroka* (Sweden), *nodderkrigge* (Denmark), *pähkinähakki* (Finland), *tannenhäher*, *nusshäher*, or *arvenhäher* (Germany, Austria), and *orekhovka* or *kedrovka* (Russia).

L. Scott Johnson et al. (1987) have described the crushing and pounding of pine seeds by Clark's Nutcracker, the activities that have given the genus *Nucifraga* both its scientific and common names. Bunch et al. (1983) have also reported on how Clark's Nutcracker handles pine seeds.

The infraspecific taxonomy used here for the Eurasian Nutcracker is that of Dement'ev et al. (1954). Details of Clark's Nutcracker's life history are taken mainly from the numerous cited papers of Balda, Vander Wall, Tomback, and Hutchins. Predatory behavior is described by Mulder et al. (1978). The sublingual pouch of Clark's Nutcracker is described in great detail by Bock et al. (1973). Pouch contents have been documented by Richmond and Knowlton (1894), Vander Wall and Balda (1977), Büchi (1955), and Turček and Kelso (1968). The study of Clark's Nutcracker's preferences among pinyon pine trees in northern Arizona is described in Christensen et al. (1991). The thermal "shells" utilized by nutcrackers to gain elevation are, in effect, packets of air that rise like free balloons, and can be transported some distance laterally. They have been implicated in the soaring flight of large birds (Cone 1962), and it has been speculated that they are responsible for long-distance movement of masses of tree pollen (Lanner 1966). Hutchins' results are reported in Hutchins and Lanner (1982), and Mattes's in Mattes (1982).

Data on cache sizes are from Hutchins and Lanner (1982), Mattes (1982), Turček and Kelso (1968), Saito (1983b), and Dimmick (1993). Details of the magnitude of caching of pine nuts by nutcrackers can be found in Vander Wall and Balda (1977), Tomback (1978), Hutchins and Lanner (1982), Dimmick (1993), and Mattes (1982).

An exception to the obsessive caching behavior of Clark's Nutcrackers has been reported in juveniles:

> Adults harvested and cached seeds intensively. They were efficient. . . . Juveniles took a great deal of time, played around, and mostly ate seeds instead of making seed stores for later use. . . . If they did cache, they immediately recovered seeds in a nonsense cache sequence. (Dimmick 1993, p. 106)

By a "nonsense cache" Dimmick means caches of seeds or non-food items (bark chunks, pollen cones, pebbles) that are repeatedly cached and recovered, as though in play.

Vander Wall (1990) suggests that the caching of seeds on harsh ridgetop sites, and where a deep snowpack will accumulate, can be beneficial to nutcrackers as rodent-avoidance measures. Thus seeds are moved from the forest, where rodent population densities are high, to environments of low rodent density. Detailed studies of rodent-nutcracker interaction await execution.

Some sources on Japanese stone pine and the Japanese nutcracker are Saito 1983a and 1983b.

CHAPTER 6

Crocq's work is described in his book on the Eurasian Nutcracker (1990) and in his doctoral thesis of 1978. The information on olfaction by corvids and other birds is from Buitron and Nuechterlein (1985). According to these authors, birds with a well-developed olfactory bulb often use olfaction to find food. Corvids do not have well-developed olfactory bulbs. J. Stokley Ligon's comments are recorded in his *New Mexico Birds* (1961). The experiments described here are reported in Vander Wall (1982).

Experiments conducted by Dimmick (1993) showed that juvenile Clark's Nutcrackers—first-year birds—were less accurate in locating and recovering their caches than were adults; and that yearlings were intermediate in their performance. He concludes that spatial memory has a strongly inherited component, but that it also improves with experience.

For a critical and wide-ranging discussion of cache recovery see Vander Wall (1990). See also Balda (1980), Kamil and Balda (1990) and Balda and Kamil (1989, 1992) for details of behavioral studies. Indirect evidence of memory-use has been reported by Swanberg (1951), Mezhenny (1964), Crocq (1978, 1990), and Tomback (1980).

CHAPTER 7

The information on planting *P. sibirica* in Finland is from Holzer (1972), Stefansson (1955), and conversations with Finnish colleagues, especially Teijo Nikkanen, at the Punkaharju Research Station. Mature Siberian stone pines can also be found in public parks and gardens in Helsinki. Risto Vaisanen of the University of Helsinki has described for me the influx of Siberian Nutcrackers into Finland in 1968. Details of past eruptions of Eurasian Nutcrackers can be found in Dement'ev et al. (1954) and Crocq (1990) for the Siberian Nutcracker; and of Clark's Nutcrackers in Davis and Williams (1957, 1964), Westcott (1964), Fisher and Myres (1979), and Vander Wall et al. (1981).

The European Nutcracker apparently erupts rarely, perhaps because it is strongly territorial (Swanberg 1951). Dement'ev et al. (1954) record only one such instance, in 1885.

Conrads and Balda (1979) studied the behavior of Siberian Nutcrackers in Bielefeld, Germany following the moderate eruption of 1977. After subsisting largely on hazelnuts in and around the city, these birds did poorly on seeds of Austrian pine and Japanese larch. They refused the seeds of Norway and Serbian spruces, but some of them, when presented with Siberian stone pine seeds at chosen feeders, ate and cached them "almost to the total exclusion of other seeds." Only those birds that were fed stone pine seeds survived. In later experiments in which they supplied large quantities of stone pine seeds, Balda and Conrads (1990) determined that single birds cached 55,000 to 181,000 seeds during an intensive 40-day caching period. These birds displayed an accuracy rate of 84 percent when recovering seeds from their caches.

Wästljung (1989) observed that European Nutcrackers preferred to harvest hazelnuts from bushes growing in the open, while red squirrels preferred to harvest from bushes growing under a forest canopy. At Punkaharju, resident Siberian Nutcrackers routinely feed also on the winged seeds of the Balkan white pine, and have effected its regeneration in and around the plantations.

Do nutcrackers that have migrated out of a foodless area return to the same forests? Vander Wall et al. (1981) noted a northward movement of birds in the summer of 1978, which they presumed to be the same population seen flying south in fall 1977, but they did not know whence the birds had come or where they were going. There appears to be no evidence of such "returns of the natives" beyond mere presumption. There is a great need for studies of nutcracker population biology.

CHAPTER 8

The question whether pines could be dependent on nutcrackers for their regeneration has not preoccupied most other researchers as it has me. Of the Europeans, only Crocq (1990) appears to have considered the question important. He has argued against an early suggestion by Holtmeier (1966) that the downslope rolling of fallen cones could regenerate Swiss stone pine from seeds shaken loose and lodged in the soil. Crocq also mentions never having seen seedlings emerge from seeds within fallen cones, and having searched to no avail for seedlings among 150 year-old cone-bearing stone pines planted outside the range of nutcrackers. Thus Crocq concludes that Swiss stone pine indeed relies on the European Nutcracker for its establishment.

Dispersal of *Caesaria*, one-seed juniper, and blackberry seeds has been reported upon by Howe and Vande Kerckhove (1979), Salomonson (1978), and Jordano (1982) respectively. The Short-Tailed Fruit Bat data are from Fleming (1988). The co-evolved mutualism of the Dodo and the Tambalacoque tree inferred by Temple (1977) has been challenged by Owadally (1979). Theirs is a scientific controversy that can never be resolved because its main party has been extinct for three centuries.

Howe (1984) argues persuasively that "obligate one-to-one mutualisms between species pairs are rare in practice and anomalous in theory." But those of his arguments that draw upon data regarding seed dispersal by birds relate entirely to plants with fleshy fruits. The development of further mutualism theory risks being trivial if it ignores corvid-pine mutualisms.

Hutchins's combination of intensive observation of all vertebrates visiting tree crowns, and quantification of those animals' activities has not yet been duplicated in other corvid-pine mutualisms. Only studies of this kind can definitively show a dependency of a pine species on a single corvid. Similar studies are needed to measure the dependence on nutcrackers of Siberian, Japanese, and Korean stone pines. The fauna of the Alps may have been modified to the point where detailed studies of this nature might not be relevant to regeneration of Swiss stone pine in the wild.

The persistence of the rotting cone hypothesis is shown by its appearance in Hosie (1990), a publication of the Canadian Forestry Service.

An early observation of whitebark pines revegetating a burned area was made by J. V. Hofmann (1917) around Mount Adams, Washington. Not knowing of Clark's Nutcracker's role, Hofmann assumed that clumped seedlings over a mile from a seed source resulted from rodent caches made just prior to an 1892 fire. This led him to think whitebark pine seeds could remain dormant at least twenty-one years, and were "remarkable examples of delayed germination."

CHAPTER 9

The calculation of how many oat seedlings are needed to produce a gram of IAA is from Salisbury and Ross (1969). The estimate of how many Clark's Nutcrackers inhabited the Squaw Basin area is based on Harry Hutchins' comment that fifty birds is the most he ever counted in Squaw Basin, and his report of a 150-bird flock at his Mt. Washburn study area in Yellowstone National Park. The use of tree diameters as an approximation of age distribution is not precise, but it alleviates drudgery in working with large numbers of trees (Buchholz and Pickering 1978). Whether Engelmann spruce was able to invade the whitebark pine groves because of the pines moderating the environment there would be contested by some ecologists on the basis of theory (Peet and Christensen 1980). More detailed studies are needed to estimate the effectiveness and the timing of Clark's Nutcrackers in restocking burned areas. A paper by Tomback et al. (1990) examined forest cover on areas burned by large fires about twenty-six years previously. Whitebark pine had started to "return" within five years, shortly after some of the wind-dispersed conifers, and there was some evidence that nutcrackers were more effective than wind in transporting seeds deep into large burns. More precise studies of burns throughout the range of whitebark pine are needed to better estimate the value of the ecosystem services performed by nutcrackers.

The study of whitebark pine genetic structure was made by Furnier et al. (1987), utilizing starch gel electrophoresis to assay nine enzyme systems representing eleven loci (genes) in seed endosperm tissue. Their material came from 110 trees from fifty-two clumps in two widely separated populations, and constitutes the most exhaustive genetic analysis to date of whitebark pine. They found that in twenty-three of the thirty-five clumps in which more than one trunk's seeds were sampled, there were at least two genetically distinct individuals. In a smaller-scale study utilizing only two to four loci, Linhart and Tomback (1985) found two or more genetically distinct trunks in twenty of twenty-five clumps of whitebark and limber pines. Schuster and Mitton (1991), however, found that in a limber pine population growing on an isolated butte east of the Rocky Mountains, eighty-seven of 108 apparent tree clumps contained only genetically similar individuals. It is unclear why these results differ so markedly from those of others; but the authors suggest they may have confused damaged trees that assumed a multi-trunked habit with clumps of multiple individuals.

Root-grafting can be considered almost ubiquitous among forest trees of the same species (Graham and Bormann 1966) and occasionally between closely related species. Transfer of radioisotopes (Yli-Vaakuri 1954, Kuntz and Riker 1956), phytocides (Bormann and Graham 1960), dyes (Yli-Vaakuri 1954), and fungal spores from roots of one tree to another has been established, and the longevity of living stumps demonstrates that hormones and carbohydrates also can be shunted from tree to tree through grafts (Lanner 1961, Wold and Lanner 1965). Since root-grafting occurs routinely between trees much further apart than those in clumps (Lanner 1961), and since there is no evidence whatever that growing in clumps enhances a tree's fitness, it is premature to speculate on the presumed evolutionary advantages of growing in root-grafted clumps (Tomback and Linhart 1990), a role for grafting in kin selection (Schuster and Mitton 1991), or the transmission of cone-crop coordination signals through grafts (Smith and Balda 1979)!

Another result of purposeful seed caching by nutcrackers is to create mixed forests. Clark's Nutcrackers cause whitebark pine seedlings to appear in clonal stands of quaking aspen and beneath aging overstories of bark beetle-riddled lodgepole pine. This aspect of seed caching, and that of forest age-structure, warrant detailed attention by forest ecologists.

CHAPTER 10

If bears were like other mammals, the long period of inactivity during hibernation would result in osteoporosis due to loss of calcium from the bones. But because a hibernating bear does not urinate, it cannot eliminate calcium. Thus denning bears do not develop osteoporosis, and even continue to form bone during hibernation (Floyd 1990). For further details on the physiology of bear hibernation see Nelson (1973, 1979). According to Jonkel (1967) whitebark pine nuts contain an estrogenic substance that suppresses reproduction. According to Harold D. Picton of Montana State University, however, there is insufficient evidence on the matter to draw conclusions.

The great diversity of the Grizzly Bear diet is detailed by Mattson et al. (1991). They point out that dietary trends reflect long-term variability in food resources, and that bear density is limited largely by the availability of berries and whitebark pine nuts. Apparently bears sometimes raid caches of limber pine cones (Craighead 1979, Mattson and Jonkel 1990). According to Mattson and Jonkel (1990) black bears "consume seeds in the canopy, or . . . break limbs off and subsequently consume the seeds on the forest floor." Black bears in Montana often lost the hair from their front legs after feeding on pine nuts. The accumulated layer of pine pitch took the hair with it on peeling off. Heinrich (1989) writes that when Black Bears in the Maine woods climb to harvest beechnuts "the beech trees show fresh bear claw marks, [and] their branches are broken like the apple trees'." David Mattson says Black Bears are adept at pulling on branches without breaking them, which may explain the scarceness of whitebark pine trees with crowns injured by foraging.

In addition to climbing trees and raiding squirrel caches, bears can also get whitebark pine seeds by foraging on cones that fall to the ground. According to Mattson et al (1991), many such "overwintered" seeds were available in Yellowstone in the spring and summer of 1990. These resulted from the huge cone crop of 1989, which was apparently more than the squirrels and nutcrackers could remove from the trees.

Russian scientists have documented increased conflict between Brown Bears and humans during years when the Siberian stone pine has poor seed crops (Mattson and Jonkel 1990).

CHAPTER 11

Food use of limber pine nuts has received little attention. David Hurst Thomas (American Museum of National History) argues (personal correspondence) that the large number of grinding stones at Alta Toquima Village (a high-elevation prehistoric site in central Nevada) can only be accounted for by use of the nuts as food. Physical evidence is, however, not abundant. Limber pines grow on the Toquima Mountains site today. I am told by Robert Bettinger (University of California, Davis) that the Northern Shoshones ate limber pine seeds. These nuts were also collected and eaten by early settlers in southeast Idaho's Bear Lake Valley, according to Darrel R. Bienz.

Salish use of whitebark pine nuts (as well as the trees' phloem) has been documented by Turner (1988). There is surprisingly little accessible information on the use and commerce of Swiss stone pine nuts, other than word of mouth. There is however an abundance of material, nearly all in Russian, on *kedr* seed use. Levin and Potapov (1964) cover several of the Siberian ethnic groups. Other details were gleaned from Lebedeva and Saf'yanova (1979), Ivanov (1936), and articles in Shimanyuk (1963). According to Stubbe (1973), the harvesting of Siberian stone pine nuts is very difficult work, but the pay was good enough to lure factory workers and even medical doctors into taking their vacations to engage in it. A vast amount of data on Siberian stone pine seeds can be found in Shimanyuk (1963). Ethnographic information on Japanese stone pine is mostly in Russian, as the species is primarily Siberian in distribution. I have consulted Tikhomirov (1936), and Tugolukov (1959) on uses of Japanese stone pine.

CHAPTER 12

Further details on the relationship of nutcrackers and Great Basin bristlecone pine appear in Lanner et al. (1984) and Lanner (1988). Florence Merriam Bailey (1928) reported Clark's Nutcrackers harvesting bristlecone pine seeds in northern New Mexico.

Knee-high bushes of limber pines grow in the Patriarch Grove, apparently from seeds brought there by nutcrackers. I have observed the caching by a nutcracker of singleleaf pinyon seeds in the Patriarch Grove. I have seen no seedlings or trees of this species there, which suggests the site exceeds the tree species' elevational tolerance.

Clevenger (1991) states there are about thirty-five to forty thousand Brown Bears in Eurasia; and thirty thousand Grizzlies in North America.

The use of ponderosa pine seed by Clark's Nutcracker was documented by Giuntoli and Mewaldt in 1978. Tomback (1978) reported its feeding on Jeffrey pine. The role of the Yellow Pine Chipmunk in regenerating Jeffrey pine is described in Vander Wall (1991, 1992).

Harry and Sue Hutchins's observations on Korean stone pine biology will be published in *Oecolgia* in 1996. The Tian Shan spruce-nutcracker interaction was described by Kirikov (1936) and translated by Leon Kelso. Shtil'mark (1963) is a good source of information on the Siberian Chipmunk.

According to Turček (1966) European Nutcrackers also locate Swiss stone pine seeds that have lain dormant in the soil for a year or more by digging around germinating seeds.

Swiss researchers Ernst Frehner and Walter Shönenberger (1994) attribute persistent dormancy in Swiss stone pine seeds to underdeveloped embryos. Seeds that take two to three years to germinate have embryos less than three millimeters long when they are picked, while in those ready to germinate the embryos are seven to nine millimeters in length. Evolution of dormancy is treated in greater detail in Lanner and Gilbert (1994).

CHAPTER 13

The *Pinus ayacahuite-strobiformis-flexilis* scenario presented here is basically that suggested in Lanner (1980), with more recent supporting evidence. The contrast between "conventional white pine" and bird pine morphology is taken from Lanner (1990b). The relative youth of limber pine as a species is suggested also by the as yet unfixed winglessness of the seed. At least as far south as Utah and as far north as Montana, one can find trees whose seeds bear vestigial wings that are ineffective as agents of wind dispersal (Lanner 1985). The suggestion has been made by Tomback (1979) and Tomback and Linhart (1990) that loss of seed wings conserves energy for the tree no longer needing to make them. Pavlik (1979), however, felt that the energy savings would be trivial, and suggested that reduction of the seed wing may have been effected by dispersers selecting seeds more easily accommodated in the mouth. In addition, one could argue that since the wing tissue is part of the cone-scale surface, it would be present whether shaped into a wing or not. Therefore the energetic cost of the wing is incurred whether the wing separates from the cone scale or not. The difficulty experienced by a corvid harvesting seeds from the cone of a conventional white pine is illustrated by this account of a Steller's Jay in a sugar pine: "In a typical assault a jay, launching itself from a nearby branch struck an open cone repeatedly until a seed was dislodged. Then, swooping down, it seized the morsel in midair, carried it to a limb, pounded the shell open, and ate the kernel." (Tevis 1953, p. 130).

Calculations made from data reported by Steinhoff and Andresen (1971) show that the average growth period of limber pine seedlings from ten of the species' northernmost locations was only thirty-eight days, while from ten southern locations it was seventy-one days. Two-year height of northern seedlings grown under controlled conditions was 44 millimeters; that of southern seedlings averaged 64 millimeters. Such growth data indicate adaptation of northern populations to low temperature and a short growing season. Artificial crosses between limber and Mexican white pines gave uncertain results (Critchfield 1986). The occurrence of white pine mutants that have entire crowns of upswept branches is illustrated by "the upright white pine," *Pinus strobus* 'Fastigiata' (Del Tredici 1993). Genetic studies of conifers have typically shown branch angle to have a high heritability relative to other traits (e.g., Snyder and Namkoong 1978). Bird pines are not the only trees whose crown form and fruit placement is adapted to the needs of a seed dispersal agent. Fruit bats, for example, chew the fruit, drink the juice, and expel the seeds and pulp of several genera of tropical trees. But their large size and weak sonar system make it impracticable for them to visit fruits that are inside the tree's foliage (Van der Pijl 1972). The "bat-fruits" they feed upon are often held away from the foliage to facilitate their visits (Whitmore 1975). The model of bird pine evolution sketched out relies on natural selection and genetic drift as mechanisms of change. Thus it is basically adaptationist and undoubtedly oversimplified. Other models are available (Tomback and Linhart 1990), but as they increase in sophistication they depart further from the available data and require more speculation.

Data on Japanese white pine are mainly from Makino (1940), Saho (1972), Li (1963), Critchfield and Little (1966), Vidaković (1991), and Hayashida (1989). *Pentaphylla* was heavily cut during World War II (Numata 1974), and according to Saho (1972) natural *himekomatsu* stands are rapidly disappearing and threatened with extinction. An obvious question is why so many bird pines have evolved among the soft pines and so few among the hard pines. Perhaps hard pine cones are less hospitable to foraging corvids because they are more likely to be prickly than those of soft pines. Or perhaps the soft pines harbor more genetic variation in such traits as seed and seed-wing size, branch angle, and seed-retention mechanisms.

For references to hybrid pinyon pines see Lanner and Phillips (1992). The confused state of pinyon pine taxonomy is brought out in Bailey (1987) and Zavarin (1988). Data on Mexican jays is from Peterson and Chalif (1973).

Arguments and counter-arguments over the issue of whitebark pine's relationship to the limber pine complex can be found in Critchfield (1986), Zavarin et al (1991), and Lanner (1990a). The molecular genetic studies that appear to have ended the argument are those of Krutovskii et al. (1990 and 1994). Krutovskii and his colleagues studied two attributes of the stone pines—genetic variation in enzyme systems (Krutovskii et al. 1994) and structure of the DNA in their chloroplasts (Krutovskii and Wagner, not yet published). Their study of the genes controlling molecular structure of sixteen to twenty enzymes (isozymes) showed that the five *Cembrae* pines are indeed closely related genetically, presumably all originating from an ancient Siberian stone pine. Whitebark pine fell squarely within the *Cembrae*. These results "have been confirmed by the recently obtained data of cp DNA restriction fragment analysis" (Krutovskii et al. 1994). Such molecular studies are far more direct than those of morphology, or even gum turpentine chemistry (Zavarin et al. 1991), because they study direct products of gene action (isozymes) or differences in actual structure of the genetic material (the DNA molecule).

Tomback and Linhart (1990) have suggested that the reasons for lack of differentiation in Clark's Nutcracker are a relatively small range in which gene flow swamps local adaptation, and eruptions in which nutcracker populations mix. But the range of Clark's Nutcracker, which is about 20 degrees of longitude by 25 degrees of latitude, can be superimposed on a map of Asia to include five subspecies of the Eurasian Nutcracker. Also, as documented in chapter 7, Eurasian Nutcrackers erupt too. Actually, differentiation of bird species into subspecies does not seem to require huge land areas. Utah ornithologist William H. Behle (1985) documents twenty-two native resident bird species represented by at least two subspecies breeding in the state. Many of these have ranges smaller or comparable in size to that of Clark's Nutcracker. Data on *Miocitta galbreathi* and its resemblance to Pinyon Jays and Clark's Nutcracker are from Brodkorb (1978 and private communication). The evolutionary scenario of adaptation by jays and nutcrackers to a diet of pine nuts can be found in Vander Wall and Balda (1981).

CHAPTER 14

Whether self-pollination of whitebark pine would indeed prove damaging is conjectural, and based on experience with numerous other pines. Siberian stone pine, however, can apparently tolerate rather high levels of inbreeding (Politov and Krutovskii 1992), and this may prove to be true of whitebark pine. Fire history data are from Agee (1993). Estimates of how a genetically variable species will respond to climatic change are largely guesswork. It is customary to assume that a plant species can tolerate the climatic conditions of its natural range only. Yet centuries of experience in the establishment of exotic plants—including most of our crop plants and numerous trees—clearly show that many species can thrive and even reproduce normally under conditions very different from those of their homeland. Unfortunately, there are also many that do not; thus predictions are risky. According to one influential view, mean annual temperatures are projected to increase by about 2.5 degrees Celsius by the middle of the twenty-first century (Schneider 1989).

See Arno (1986) for an overview of whitebark pine's problems; and Hoff and Hagle (1990) for a review of the white pine blister rust situation in western North America, and details of the resistance mechanisms found in attacked pines. Data on whitebark pine damage from the disease are from Kendall and Arno (1990) and Keane and Arno (1993). Recent developments in sugar pine's response to rust attack have been described by Kinloch and Dulitz (1990).

A conundrum that seems to have escaped notice regards whitebark pine's low level of resistance to white pine blister rust. If, as all the evidence suggests, whitebark pine has recently speciated from a Eurasian forebear, why has it not the same high resistance we see in the present-day Eurasian stone pines? *Cronartium ribicola's* relationship with five-needled pines is believed to have begun in the Jurassic, shortly after *Pinus* first appeared and split into subgenera *Strobus* and *Pinus* (Millar and Kinloch 1991).

REFERENCES

Agee, James K. 1993. *Fire Ecology of Pacific Northwest Forests.* Island Press, Washington, D.C., 493 p.

Amadon, Dean. 1944. The genera of Corvidae and their relationships. *American Museum Novitates, no.* **1251**:1–21.

Angell, Tony. 1978. *Ravens, Crows, Magpies, and Jays.* University of Washington Press, Seattle. 112 p.

Arno, Stephen F. 1986. Whitebark pine cone crops: a diminishing source of wildlife food? *Western Journal of Applied Forestry* **1**:92–94.

Bailey, D. K. 1975. *Pinus albicaulis. Curtis's Botanical Magazine* **180(3)**:141–147.

———. 1987. A study of *Pinus* subsection *Cembroides* I: the single-needle pinyons of the Californias and the Great Basin. *Notes Royal Botanical Garden Edinburgh* **44**:275–310.

Bailey, Florence Merriam. 1928. *Birds of New Mexico.* New Mexico Department of Game and Fish. Judd and Detweiler, Washington, D.C., 807 p.

Balda, Russell P. 1978. Are seed caching systems co-evolved? Paper presented at symposium on co-evolutionary systems in birds, *Seventeenth International Ornithological Congress,* Berlin, June 4–11, 1978.

———. 1980. Recovery of cached seeds by a captive *Nucifraga caryocatactes. Zeitschrift Tierpsychologie* **52**:331–346.

———. 1987. Avian impacts on pinyon-juniper woodlands. In Everett, R. L. (compiler), *Proceedings Pinyon-Juniper Conference,* Reno, Nevada, Jan. 13–16, 1986. U.S. Department of Agriculture Forest Service General Technical Report INT-215:525–533.

Balda, Russell P., and Gary C. Bateman. 1971. Flocking and annual cycle of the Piñon Jay, *Gymnorhinus cyanocephalus. Condor* **73**:287–302.

Balda, Russell P., and Klaus Conrads. 1990. Freilandbeobachtungen an Sibirischen Tannenhähern (*Nucifraga caryocatactes macrorhynchos*) 1977/78. *Bielefeld. Ber. Naturwiss. Verein Bielefeld u. Umgegend* **31**:1–31.

Balda, Russell P., and Alan C. Kamil. 1989. A comparative study of cache recovery by three corvid species. *Animal Behaviour* **38**:486–495.

———. 1992. Long-term spatial memory in Clark's nutcracker, *Nucifraga columbiana. Animal Behaviour* **44**:761–769.

Behle, William H. 1985. *Utah Birds: Geographic Distribution and Systematics.* Occas. Publ. No. 5, Utah Museum of Natural History, University of Utah, Salt Lake City, 147 p.

Benkman, Craig W., Russell P. Balda, and Christopher C. Smith. 1984. Adaptations for seed dispersal and the compromises due to seed predation in limber pine. *Ecology* **65**:632–642.

Bibikov, D. I. 1948. On the ecology of the nutcracker. *Trudy Pechoro-Ilychkogo Gosud. Zapovednik* **4**:89–112.

Bock, Walter J., Russell P. Balda, and Stephen B. Vander Wall. 1973. Morphology of the sublingual pouch and tongue musculature in Clark's Nutcracker. *Auk* **90**:491–519.

Bormann, F. H., and B. F. Graham, Jr. 1960. Translocation of silvicides through root grafts. *Journal of Forestry* **58**:402–403.

Bossema, I. 1979. Jays and oaks: an eco-ethological study of a symbiosis. *Behaviour* **70**:1–117.

Brodkorb, Pierce. 1978. Catalogue of fossil birds, Part 5 (Passeriformes). *Bulletin of Florida State Museum of Biology* **23**:139–228.

Buchholz, Kenneth, and Jerry L. Pickering. 1978. DBH-distribution analysis: an alternative to stand-age analysis. *Bulletin Torrey Botanical Club* **105**:*282–288.*

Büchi, O. 1955. La voracité du Cassenoix. *Nos Oiseaux, Neuenberg,* **23**:145–146.

Buitron, Deborah, and Gary Z. Nuechterlein. 1985. Experiments on olfactory detection of food caches by black-billed magpies. *Condor* **87**:92–95.

Bunch, Kenneth G., Gary Sullivan, and Diana F. Tomback. 1983. Seed manipulation by Clark's Nutcracker. *Condor* **85**:372–373.

Cheff, Vern E. "Bud" Jr. 1984. What has happened to the Montana Grizzly? Unpublished ms., 8 p.

Christensen, Kerry M., and Thomas G. Whitham. 1991. Indirect herbivore mediation of avian seed dispersal in pinyon pine. *Ecology* **72**:534–542.

Christensen, Kerry M., Thomas G. Whitham, and Russell P. Balda. 1991. Discrimination among pinyon pine trees by Clark's Nutcrackers: effects of cone crop size and cone characters. *Oecologia* **86**:402–407.

Clevenger, Anthony. 1991. The phantom bear of the Spanish Sierras. *Wildlife Conservation* **94**:34–45.

Cone, Clarence D., Jr. 1962. Thermal soaring of birds. *Scientific American* **50**: 180–209.

Conrads, K., and R. P. Balda. 1979. Überwinterungschancen Sibirischer Tannenhäher (*Nucifraga caryocatactes macrorhynchos*) im Invasionsgebiet. *Ber. Naturwiss. Vereins Bielefeld* **24**:115–137.

Craighead, Frank C., Jr. 1979. *Track of the Grizzly.* Sierra Club Books, San Francisco.

Critchfield, William B. 1986. Hybridization and classification of the white pines (*Pinus* section *Strobus*). *Taxon* **35**:647–656.

Critchfield, William B., and Elbert L. Little, Jr. 1966. *Geographic Distribution of Pines of the World.* U.S. Department of Agriculture Forest Service Miscellaneous Publication 991.

Crocq, Claude. 1978. Écologie du Casse-noix (*Nucifraga caryocatactes* L.) dans les Alpes françaises du sud. Thèse, l'Univ d'Aix-Marseille, 189 p.

———. 1990. *Le Casse-Noix Moucheté* (*Nucifraga caryocatactes*). Lechevalier-Chabaud.

Cutright, Paul Russell. 1969. *Lewis and Clark: Pioneering Naturalists.* University of Illinois Press, Urbana.

Darley-Hill, S., and W. C. Johnson. 1981. Acorn dispersal by blue jays (*Cyanocitta cristata*). *Oecologia* **50**:231–232.

Davis, John, and Laidlaw Williams. 1957. Irruptions of the Clark's Nutcracker in California. *Condor* **59**:297–307.

———. 1964. The 1961 irruption of the Clark's Nutcracker in California. *Wilson Bulletin* **76**:10–18.

Del Tredici, Peter. 1993. The upright white pine. *Arnoldia* **53 (1)**: 24–31.

Dement'ev, G. P., N. A. Gladkov, A. M. Sudilovskaya, E. P. Spangenberg, L. V. Boehme, I. B. Volchanetskii, M. A. Vointvenskii, N. N. Gorchakovskaya, M. N. Korelov, and A. K. Rustamov. 1954. *Birds of the Soviet Union,* Vol. 5. Published for the Smithsonian Institution and the National Science Foundation, Washington, D.C., by the Israel Program for scientific translations.

Dimmick, Curt R. 1993. Life history and the development of cache-recovery behaviors in Clark's Nutcracker. Ph.D. diss., Northern Arizona University, Flagstaff, 209 p.

Farris, G. J. 1983. California pignolia: seeds of *Pinus sabiniana. Economic Botany* **37**:201–206.

Finley, Robert B., Jr. 1969. Cone caches and middens of *Tamiasciurus* in the Rocky Mountain region. *University of Kansas Museum of Natural History, Miscellaneous Publication* **51**:223–273.

Fisher, Robert M., and M. T. Myres. 1979. A review of factors influencing extralimital occurrences of Clark's Nutcracker in Canada. *Canadian Field-Naturalist* **94**:43–51.

Fleming, T. H. 1988. *The Short-Tailed Fruit Bat. A Study in Plant-Animal Interactions.* Univ. Chicago Press, Chicago.

Floyd, Charles Timothy. 1990. Bone metabolism in black bears: potential applications to human osteoporoses. In D. W. Reed, editor, *Spirit of Enterprise, the 1990 Rolex Awards,* Buri International, Bern, pp. 24–26.

Frehner, Ernst, and Walter Schönenberger. 1994. Experiences with reproduction of cembra pine. In Schmidt, W.C., and F.-K. Holtmeier (compilers), *Proceedings—International Workshop on Subalpine Stone Pines and their Environment: the Status of Our Knowledge,* St. Moritz, Switzerland, Sept. 5–11, 1992. U.S. Department of Agriculture Forest Service General Technical Report INT-GTR-309:52–55.

Furnier, Glenn R., Peggy Knowles, Merlise A. Clyde, and Bruce P. Dancik. 1987. Effects of avian seed dispersal on the genetic structure of whitebark pine populations. *Evolution* **41**:607–612.

Giuntoli, Mervin, and L. Richard Mewaldt. 1978. Stomach contents of Clark's Nutcracker collected in western Montana. *Auk* **95**:595–598.

Goodwin, Derek. 1976. *Crows of the World.* Cornell University Press, Ithaca, N.Y.

Graham, Ben F., Jr., and F. H. Bormann. 1966. Natural root grafts. *Botanical Review* **32**:255–292.

Grinnell, J. 1936. Up-hill planters. *Condor* **38**:80–82.

Harlow, William M. 1931. The identification of the pines of the United States, native and introduced, by needle structure. New York State College of Forestry Technical Publication 32, Syracuse, N.Y., 21 p.

Harlow, William M., W. A. Coté, Jr., and A. C. Day. 1964. The opening mechanism of pine cone scales. *Journal of Forestry* **62**:538–540.

Harlow, William M., Ellwood S. Harrar, James W. Hardin, and Fred M. White. 1991. Textbook of Dendrology, 7th ed. McGraw-Hill, New York.

Hayashida, Mitsuhiro. 1989. Seed dispersal and regeneration patterns of *Pinus parviflora* var. *pentaphylla* on Mt. Apoi in Hokkaido. *Research Bulletin of College Experimental Forests, Faculty of Agriculture, Hokkaido University* **46**:177–190.

Heinrich, Bernd. 1989. *Ravens in Winter.* Summit Books, New York. 379 p.

Hendrickson, William H., and James E. Lotan. 1971. Identification of whitebark and limber pines based on needle resin ducts. *Journal of Forestry* **69**:584.

Hoff, Ray, and Susan Hagle. 1990. Diseases of whitebark pine with special emphasis on white pine blister rust, in Schmidt, W. C., and K. J. McDonald (compilers), *Proceedings—Symposium on Whitebark Pine Ecosystems: Ecology and Management of a High-Mountain Resource.* USDA Forest Service General Technical Report INT-270, pp. 179–190.

Hofmann, J. V. 1917. Natural reproduction from seed stored in the forest floor. *Journal of Agricultural Research* **XI(1)**:1–26, 7 pl.

Holtmeier, Friedrich-Karl. 1966. Die ökologische Funktion des Tannenhähers in Zirben-Lärchenwald und an der Waldgrenze des Oberengadins. *Jour. für Ornithologie (Berlin)* **107**:337–345.

Holzer, Kurt. 1972. Intrinsic qualities and growth-potential of *Pinus cembra* and *Pinus peuce* in Europe. In *Biology of Rust Resistance in Forest Trees,* Proceedings of a NATO-IUFRO Advanced Study Institute, Moscow, Idaho, Aug. 17–24, 1969. USDA Forest Service Miscellaneous Publication 1221:99–110.

Hosie, R. C. 1990. *Native Trees of Canada,* 8th ed. Published by Fitzhenry and Whiteside for Canadian Forestry Service, Markham, Ontario, 380 p.

Howe, Henry F. 1984. Constraints on the evolution of mutualisms. *American Naturalist* **123**:764–777.

———, and G. A. Vande Kerckhove. 1979. Fecundity and seed dispersal of a tropical tree. *Ecology* **60**:180–189.

Hutchins, H. E., and R. M. Lanner. 1982. The central role of Clark's Nutcracker in the dispersal and establishment of whitebark pine. *Oecologia* **55**:192–201.

Ivanov, V. 1936. Kedr v vostochnoii sibiri. *Sov. Kraevedenie* **9**:88–98.

Johnson, L. Scott, John M. Marzluff, and Russell P. Balda. 1987. Handling of pinyon pine seed by Clark's Nutcracker. *Condor* **89**:117–125.

Johnson, W. C., and C. S. Adkisson. 1985. Dispersal of beech nuts by blue jays in fragmented landscapes. *American Midland Naturalist* **113**:319–324.

———. 1986. Airlifting the oaks. *Natural History* **95**:40–47.

Johnson, W. C., and Thompson Webb, III. 1989. The role of blue jays (*Cyanocitta cristata* L.) in the postglacial dispersal of fagaceous trees in eastern North America. *Journal of Biogeography* **16**:561–571.

Jonkel, Charles J. 1967. The ecology, population dynamics, and management of the Black Bear in the spruce-fir forest of northwestern Montana. Ph.D. diss., University of British Columbia, Vancouver.

Jordano, Pedro. 1982. Migrant birds are the main seed dispersers of blackberries in southern Spain. *Oikos* **38**:183–193.

Kamil, Alan C., and Russell P. Balda. 1990. Differential memory for different cache sites by Clark's Nutcrackers (*Nucifraga columbiana*). *Journal of Experimental Psychology: Animal Behavior Proceedings* **16**:162–168.

Keane, Robert E., and Stephen F. Arno. 1993. Rapid decline of whitebark pine in western Montana: evidence from 20-year remeasurements. *Western Journal of Applied Forestry* **8**:44–47.

Kendall, Katherine C. 1983. Use of pine nuts by grizzly and black bears in the Yellowstone area. *International Conference on Bear Research and Management* **5**: 166–173.

Kendall, Katharine C., and Stephen F. Arno. 1990. Whitebark pine: an important but endangered wildlife resource. In Schmidt, W. C., and K. J. McDonald (compilers), *Proceedings—Symposium on Whitebark Pine Ecosystems: Ecology and Management of a High-Mountain Resource.* USDA Forest Service General Technical Report INT-270:264–273.

Kilham, Lawrence. 1991. *The American Crow and the Common Raven.* Texas A&M University Press, College Station, Texas.

Kinloch, Bohun B., and David Dulitz. 1990. White pine blister rust at Mountain Home Demonstration State Forest: a case study of the epidemic and prospects for genetic control. USDA Forest Service Research Paper PSW-204, Berkeley, Ca., 7p.

Kirikov, S. V. 1936. On the ecological relationship between the nutcracker (*Nucifraga caryocatactes* L.) and the spruces (*Picea*). Translated by Leon Kelso. *Izvestiia Akademii Nauk USSR, Otdel. Mat. I Estest. Nauk, Ser. Biol.* **6**:1235–1250 .

Korelov, M. N. 1948. On the ecology of the kedrovka, *Nucifraga caryocatactes rothschildi* Hart. Translated by Leon Kelso. *Vestnik. Akad. Nauk Kazakhskoi SSR* **5**:72–75.

Kozhevnikov, A. M. 1963. Fruit bearing of the Siberian stone pine in the western part of the Transbaikal. In A. P. Shimanyuk, *Fruiting of the Siberian stone pine in east Siberia,* Acad. Sci. USSR Siberian Dept., Trans. Inst. Forestry and Wood Processing Vol. 2:80–97. Transl. Israel Prog. Sci. Transl. Publ. U.S. Department of Agriculture and National Science Foundation, 1966.

Krushinskaya, N. L. 1970. On the problem of memory. *Priroda* **9**:75–78.

Krutovskii, Konstantin V., Dimitri V. Politov, and Yuri P. Altukhov. 1990. Interspecific genetic differentiation of Eurasian stone pines for isoenzyme loci. *Soviet Genetics* **26 (4)**:440–451.

———. 1994. Genetic differentiation and phylogeny of stone pine species based on isozyme loci. In W. C. Schmidt, and F.-K. Holtmeier, (compilers), *Proceedings, International Workshop on Subalpine Stone Pines and their Environment: the Status of Our Knowledge,* St. Moritz, Switzerland, Sept. 5–12, 1992. U.S. Department of Agriculture Forest Service General Technical Report INT-GTR-309:19–30.

Krutovskii, Konstantin V., and D. B. Wagner. 1996. Phylogeny of *Pinus albicaulis,* the only North American stone pine—a chloroplast DNA study. *Forest Genetics* (in press).

Kuntz, J. E., and A. J. Riker. 1956. The use of radioisotopes to ascertain the role of root grafting in the translocation of water, nutrients, and disease-inducing organisms among forest trees. *Proceedings of the International Conference on Peaceful Uses of Atomic Energy (Geneva),* **12**:144–148.

Lanner, Ronald M. 1961. Living stumps in the Sierra Nevada. *Ecology* **42**:170–173.

———. 1966. Needed: a new approach to the study of pollen dispersion. *Silvae Genetica* **15**:50–52.

———. 1974. A new pine from Baja California and the hybrid origin of *Pinus quadrifolia. Southwestern Naturalist* **19(1)**:75–95.

———. 1980. Avian seed dispersal as a factor in the ecology and evolution of limber and whitebark pines. In B. P. Dancik and K. O. Higginbotham, eds., *Sixth North American Forest Biology Workshop Proc.* Univ. of Alberta, Edmonton, Aug. 11–13, 1980: 15–48.

———. 1981. *The Piñon Pine, A Natural and Cultural History.* University of Nevada Press, Reno.

———. 1982. Adaptations of whitebark pine for seed dispersal by Clark's Nutcracker. *Canadian Journal of Forest Research* **12**:391–402.

———. 1985. Effectiveness of the seed wing of *Pinus flexilis* in wind dispersal. *Great Basin Naturalist* **45(2)**:318–320.

———. 1988. Dependence of Great Basin bristlecone pine on Clark's Nutcracker for regeneration at high elevations. *Arctic and Alpine Research* **20**:358–362.

———. 1990a. Biology, taxonomy, evolution, and geography of stone pines of the world. In W. C. Schmidt and K. J. McDonald (compilers), *Proceedings, Symposium on Whitebark Pine Ecosystems: Ecology and Management of a High-Mountain Resource,* Bozeman, Mont., March 29–31, 1989. U.S. Department of Agriculture Forest Service General Technical Report INT-270:14–24.

———. 1990b. Morphological differences between wind-dispersed and bird-dispersed pines of subgenus *Strobus*. In W. C. Schmidt and K. J. McDonald (compilers), *Proceedings-Symposium on Whitebark Pine Ecosystems: Ecology and Management of a High-Mountain Resource,* Bozeman, Mont., March 29–31, 1989.

U.S. Department of Agriculture Forest Service General Technical Report INT-270: 371–372.

Lanner, Ronald M., and Barrie K. Gilbert. 1994. Nutritive value of whitebark pine seeds, and the question of their variable dormancy. In W. C. Schmidt and F.-K. Holtmeier (compilers), *Proceedings, International Workshop on Subalpine Stone Pines and their Environment: the Status of Our Knowledge,* St. Moritz, Switzerland, Sept. 5–11, 1992. U.S. Department of Agriculture Forest Service General Technical Report INT-GTR-309:206–211.

Lanner, Ronald M., Harry E. Hutchins, and Harriette A. Lanner. 1984. Bristlecone pine and Clark's Nutcracker: probable interaction in the White Mountains, California. *Great Basin Naturalist* **44**:357–360.

Lanner, Ronald M., and Arthur M. Phillips III. 1992. Natural hybridization and introgression of pinyon pines in northwestern Arizona. *International Journal of Plant Sciences* **153**:250–257.

Lanner, Ronald M., and Stephen B. Vander Wall. 1980. Dispersal of limber pine seed by Clark's Nutcracker. *Journal of Forestry* **78**:637–639.

Lebedeva, A. A., and A. V. Saf'yanova. 1979. [The cedar nut trade in Siberia.] Translated by Gail Fondahl. *Soviet Etnografiya* **4**:107–117.

Levin, M. G., and L. P. Potapov, editors. 1964. *The Peoples of Siberia.* University of Chicago Press, Chicago.

Li, Hui-Lin. 1963. *Woody Flora of Taiwan.* Livingston Publ. Co., Narberth, Penn., 974 p.

Ligon, J. David. 1971. Late summer-autumnal breeding of the Piñon Jay in New Mexico. *Condor* **73**:147–153.

———. 1974. Green cones of the piñon pine stimulate late summer breeding in the Piñon Jay. *Nature* **250**:80–82.

———. 1978. Reproductive interdependence of Piñon Jays and piñon pines. *Ecological Monograph* **48**:111–126.

Ligon, J. David, and D. J. Martin. 1974. Piñon seed assessment by the Piñon Jay, *Gymnorhinus cyanocephalus. Animal Behaviour* **22**:421–429.

Ligon, J. David, and J. L. White, 1974. Molt and its timing in the Piñon Jay, *Gymnorhinus cyanocephalus. Condor* **76**:274–287.

Ligon, J. Stokley. 1961. *New Mexico Birds.* University of New Mexico Press, Albuquerque.

Linhart, Yan B., and Diana F. Tomback. 1985. Seed dispersal by nutcrackers causes multi-trunk growth form in pines. *Oecologia (Berlin)* **67**:107–110.

Little, Elbert L., Jr., and William B. Critchfield. 1969. Subdivisions of the genus *Pinus* (Pines), U.S. Department of Agriculture Forest Service Miscellaneous Publication 1144.

Makino, Tomitaro. 1940. *An Illustrated Flora of Nippon, with the Cultivated and Naturalized Plants.* Hokuryu Kan, Tokyo.

Marzluff, John N., and Russell P. Balda. 1992. *The Pinyon Jay: Behavioral Ecology of a Colonial and Cooperative Corvid.* T. & A. D. Poyser, London.

Mattes, Hermann. 1982. Die Lebensgemeinschaft von Tannenhäher und Arve. Swiss Federal Inst. of Forestry Research, CH 8903 Birmensdorf, Report No. 241, 74 p.

Mattson, David J., and Charles Jonkel. 1990. Stone pines and bears. In W. C. Schmidt and K. J. McDonald (compilers), *Proceedings, Symposium on whitebark pine ecosystems: ecology and management of a high-mountain resource,* Bozeman, Mont., March 29–31, 1989. U.S. Department of Agriculture Forest Service General Technical Report INT-270:223–236.

Mattson, David J., Bonnie M. Blanchard, and Richard R. Knight. 1991. Food habits of Yellowstone grizzly bears, 1977–1987. *Canadian Journal of Zoology* **69**:1619–1629.

Mattson, David J., Daniel P. Reinhart, and Bonnie M. Blanchard. 1994. Variation in production and bear use of whitebark pine seeds in the Yellowstone area. In D. G. Despain, and P. Schullery, (editors), Proceedings, Plants and their Environment: 1st Biennial Scientific Conference on the Greater Yellowstone Ecosystem. Department of Interior, National Park Service: 103-118.

McCarthy, M. A., and R. H. Matthews. 1984. Composition of foods: nut and seed products. U.S. Department of Agriculture, Agriculture Handbook 8–12.

McCune, Bruce. 1988. Ecological diversity in North American pines. *American Journal of Botany* **75**:353–368.

McMaster, Gregory S., and Paul H. Zedler. 1981. Delayed seed dispersal in *Pinus torreyana* (Torrey Pine). *Oecologia (Berlin)* **51**:62–66.

Mezhenny, A. A. 1964. Biology of the nutcracker *Nucifraga caryocatactes macrorhynchos* in south Yakutia. Translated by Leon Kelso. *Zool. Zhurnal* **43**:1679–1687.

Millar, C. I. 1993. Impact of the Eocene on the evolution of *Pinus* L. *Annals of Missouri Botanical Garden* **80**:471–498.

———, and B. B. Kinloch. 1991. Taxonomy, phylogeny, and coevolution of pines and their stem rusts. In Hiratsuka, Y., J. K. Samoil, P. V. Blenis, P. E. Crane, and B. L. Laishley, eds. *Rusts of pine.* Proceedings of the IUFRO Rusts of Pine Working Party Conference, Sept. 18–22, 1989, Banff, Alberta, Canada. Forestry Canada Northwest Region, Northern Forestry Centre, Edmonton, Alberta. Information Report NOR-X-317:1–38.

Mirov, Nicholas T. 1967. *The Genus Pinus.* Ronald Press, New York.

Mirov, Nicholas T., and Jean Hasbrouck. 1976. *The Story of Pines.* Indiana University Press, Bloomington.

Mulder, Barry S., Brian B. Schultz, and Paul W. Sherman. 1978. Predation on vertebrates by Clark's Nutcrackers. *Condor* **80**:449–451.

Myers, Norman, 1986. Tropical deforestation and a mega-extinction spasm, in M. E. Soulé, editor, *Conservation Biology, The Science of Scarcity and Diversity.* Sinauer Associates, Sunderland, Mass., 394–409.

Neilson, James A. 1926. Bird notes from Wheatland, Wyoming. *Condor* **28**:99–102.

Nelson, Ralph A. 1973. Winter sleep in the black bear—a physiologic and metabolic marvel. *Mayo Clinic Proceedings* **48**:733–737.

———. 1979. Protein and fat metabolism in hibernating bears. *Federation Proceedings* **39**:2955–2958.

Newberry, J. S. 1857. Report upon the botany of the route. In Reports on explorations and surveys to ascertain the most practicable and economic route for a railroad from the Mississippi River to the Pacific Ocean. 33rd Congress, 2nd session Executive Document, Washington. **6(91):** 44–45.

Numata, M. 1974. *The Flora and Vegetation of Japan.* Kodansha Ltd., Tokyo.

Oring, L. W., and R. W. Seabloom. 1971. Status of Clark's Nutcracker in North Dakota. *Prairie Naturalist* **3**:55–56.

Owadally, A. W. 1979. The Dodo and the Tambalacoque tree. *Science* **203**: 1363–1364.

Pavlik, Bruce M. 1979. Comment on seed dispersal in *Pinus. Waucoba News* (Bishop Museum and Historical Society, Bishop, Calif.) III (Spring):2.

Peet, Robert K., and Norman L. Christensen. 1980. Succession: a population process. *Vegetatio* **43**:131–140.

Perry, Jesse P., Jr. 1991. *The Pines of Mexico and Central America.* Timber Press, Portland, Ore., 231 p.

Peterson, Roger Tory, and Edward L. Chalif. 1973. *A Field Guide to Mexican Birds.* Houghton-Mifflin Co., Boston, 298 p.

Politov, Dmitri V., and Konstantin V. Krutovskii. 1992. Allozyme polymorphism, heterozygosity, and mating system of stone pines. In W. C. Schmidt and F.-K. Holtmeier (compilers), *Proceedings, International Workshop on Subalpine Stone Pines and their Environment: the Status of Our Knowledge,* St. Moritz, Switzerland, Sept. 5–12, 1992. U.S. Department of Agriculture Forest Service General Technical Report INT-GTR-309:36–42.

Potter, L. D., and D. L. Green. 1964. Ecology of a northeastern outlying stand of *Pinus flexilis. Ecology* **45**:866–868.

Reimers, N. F. 1953. The food of the nutcracker and its role in the dispersal of the cedar-pine in the mountains of Khamar-Daban. Translated by Leon Kelso. *Lesnoe Khozyaistvo* **1**:63–64.

Richmond, C. W., and F. H. Knowlton. 1894. Birds of south-central Montana. *Auk* **11**:298–308.

Rikli, M. 1909. *Die Arve in der Schweiz.* Denkschrift d. Schweizer naturf. Gesellschaft 44, 455 p.

Saho, Haruyoshi. 1972. White pines of Japan. In *Biology of Rust Resistance in Forest Trees*: Proceedings of a NATO-IUFRO Advanced Study Institute, Moscow, Idaho, Aug. 17–24, 1969. USDA Forest Service Miscellaneous Publications 1221:179–199.

Saito, Shin-ichiro. 1983a. On relations of the caching by animals on the seed germination of Japanese stone pine, *Pinus pumila* Regel. *Bulletin of Shiretoko Museum (Hokkaido)* **5**:23–40.

———. 1983b. Caching of Japanese stone pine seeds by nutcrackers at the Shiretoko Peninsula, Hokkaido. *Tori* **32**:13–20.

Salisbury, Frank B., and Cleon Ross. 1969. *Plant Physiology.* Wadsworth Publ. Co., Belmont, Calif.

Salomonson, Michael G. 1978. Adaptations for animal dispersal of one-seed juniper seeds. *Oecologia* **32**:333–339.

Sargent, Charles Sprague. 1897. *The Silva of North America, a description of the trees which grow naturally in North America exclusive of Mexico.* Vol. 11 *Coniferae* (*Pinus*). Houghton, Mifflin and Co., Boston and New York.

Sauer, Martin. 1802. An account of a geographical and astronomical expedition to the northern parts of Russia, etc. Performed, by command of Her Imperial Majesty Catherine the Second, Empress of all the Russias, by Commodore Joseph Billings in the years 1785 & C. to 1794. London, Cadell and Davies. Reprinted 1972 by Richmond Publ. Co., Richmond, Surrey.

Schneider, Stephen H. 1989. The greenhouse effect: science and policy. *Science* **243**:771–781.

Schuster, William S. F., and Jeffry B. Mitton. 1991. Relatedness within clusters of a bird-dispersed pine and the potential for kin interactions. *Heredity* **67**:41–48.

Shaw, George Russell. 1914. *The genus Pinus.* Riverside Press, Cambridge, Mass.

Shimanyuk, A. P., editor. 1963. Fruiting of the Siberian stone pine in east Siberia. Acad. Sci. USSR Siberian Dept., Trans. Inst. Forestry and Wood Processing Vol. 62:80–97. Israel. Prog. Sci. Transl. Publ. U.S. Department of Agriculture and National Science Foundation 1966.

Shtil'mark, F. R. 1963. Ecology of the chipmunk (*Eutamias sibiricus* Laxm.) in cedar forests of the Western Sayan Mountains. Translated by Leon Kelso. *Zool. Zhur.* **42**:92–102.

Smith, Christopher C. 1970. The coevolution of pine squirrels (*Tamiasciurus*) and conifers. *Ecological Monograph* **40**:349–371.

Smith, Christopher C., and Russell P. Balda. 1979. Competition among insects, birds, and mammals for conifer seeds. *American Zoologist* **19**:1065–1083.

Snethen, Karen L. 1980. Whitebark pine (*Pinus albicaulis* Engelm.) invasion of a subalpine meadow. M.S. thesis, Utah State University, Logan, 76 p.

Snyder, E. B., and Gene Namkoong. 1978. Inheritance in a diallel crossing experiment with longleaf pine. USDA Forest Service Research Paper SO-140, Southern Forest Experiment Station, New Orleans, La.

Stefansson, E. 1955. Främmande barrträd i norrlandskt skogsbruk. *Svensk Papperstidning* **58**:1–11.

Stegmann, B. K. 1934. On the phylogeny of the nutcracker (Kedrovka). Translated by Leon Kelso. *Dokl. Akad. Nauk. USSR 2(2d Ser.)* **No. 4**:267–269.

Steinhoff, R. J., and J. W. Andresen. 1971. Geographic variation in *Pinus flexilis* and *Pinus strobiformis* and its bearing on their taxonomic status. *Silvae Genetica* **20**:159–167.

Stubbe, C. 1973. Die Sibirische Kiefer, die Zeder sibiriens. *Sozialistische Forstwirtschaft* **23**:123–214.

Sudworth, George B. 1908. *Forest Trees of the Pacific Slope.* Forest Service, Washington, D.C., 441 p.

Swanberg, P. O. 1951. Food storage, territory, and song in the thick-billed nutcracker. *Proceedings of the Tenth International Ornithological Conference* pp. 545–554.

Temple, Stanley A. 1977. Plant-animal mutualism: coevolution with Dodo leads to near extinction of plant. *Science* **197**:885–886.

Tevis, Lloyd, Jr. 1953. Effect of vertebrate animals on seed crop of sugar pine. *Journal of Wildlife Management* **17**:128–131.

Tikhomirov, B. A. 1936. Kedrov'iy stlannik. Translated by Gail Fondahl. *Sovietskaya Arktika* **5**:100–105.

Tomback, Diana F. 1978. Foraging strategies of Clark's Nutcracker. *Living Bird, 16th Annual,* **1977**:123–161.

———. 1979. White-bark and limber pines. *Waucoba News* (Bishop Museum and Historical Society, Bishop, Calif.) III (Winter):1.

———. 1980. How nutcrackers find their seed stores. *Condor* **82**:10–19.

———. 1982. Dispersal of whitebark pine seeds by Clark's Nutcracker: a mutualism hypothesis. *Journal of Animal Ecology* **51**:451–467.

Tomback, Diana F., and Yan B. Linhart. 1990. The evolution of bird-dispersed pines. *Evolutionary Ecology* **4**:185–219.

Tomback, Diana F., Lyn A. Hoffmann, and Sharren K. Sund. 1990. Coevolution of whitebark pine and nutcrackers: implications for forest regeneration. In W. C. Schmidt and K. J. Mc Donald (compilers), *Proceedings, Symposium on Whitebark Pine Ecosystems: Ecology and Management of a High-Mountain Resource,* Bozeman, Mont., March 29–31, U.S. Department Agriculture Forest Service General Technical Report INT- 270:118–129.

Tugolukov, V. A. 1959. [Economic life of the Okhotsk Evenki in the past and present.]. Translated by Gail Fondahl. Kratkie soobshchenie Instituta etnografii 31:46–54.

Turček, Frantisek J. 1966. Über das Wiederauffinden von im Boden versteckten Samen durch Tannen- und Eichelhäher. *Waldhygiene* **6**:215–217.

Turček, Frantisek J., and Leon Kelso. 1968. Ecological aspects of food transportation and storage in the Corvidae. *Communications in Behavior Biology, Part A,* **1**:277–297.

Turner, Nancy. 1988. Ethnobotany of coniferous trees in Thompson and Lilooet Interior Salish of British Columbia. *Economic Botany* **42**:177–194.

USDA Forest Service. 1974. Seeds of woody plants in the United States. U.S. Department Agriculture, Agriculture Handbook 450. Washington, D.C.

Van der Pijl, L. 1972. *Principles of Dispersal in Higher Plants,* 2nd ed., Springer-Verlag, New York, 162p.

Vander Wall, Stephen B. 1982. An experimental analysis of cache recovery in Clark's Nutcracker. *Animal Behaviour* **30**:84–94.

———. 1988. Foraging of Clark's Nutcrackers on rapidly changing pine seed resources. *Condor* **90**:621–631.

———. 1990. *Food Hoarding in Animals.* University of Chicago Press, Chicago, 443 p.

———. 1991. Mechanisms of cache recovery by yellow pine chipmunks. *Animal Behaviour* **41**:851–863.

———. 1992. Establishment of Jeffrey pine seedlings from animal caches. *Western Journal of Applied Forestry* **7**:14–20.

Vander Wall, Stephen B., and Russell P. Balda. 1977. Coadaptations of the Clark's Nutcracker and the piñon pine for efficient seed harvest and dispersal. *Ecological Monographs* **47**:89–111.

———. 1981. Ecology and evolution of food-storage behavior in conifer-seed-caching Corvids. *Zeitschrift Tierpsychologie* **56**:217–242.

Vander Wall, Stephen B., and Harry E. Hutchins. 1983. Dependence of Clark's Nutcracker, *Nucifraga columbiana*, on conifer seeds during the postfledging period. *Canadian Field-Naturalist* **97**:208–214.

Vander Wall, Stephen B., Stephen W. Hoffman, and Wayne K. Potts. 1981. Emigration behavior of Clark's Nutcracker. *Condor* **83**:162–170.

Vidaković, Mirko. 1991. *Conifers, Morphology and Variation.* Grafički Zavod Hrvatske, 754 p.

Wästljung, Urban. 1989. Effects of crop size and stand size on seed removal by vertebrates in hazel, *Corylus avellana. Oikos* **54**:178–184.

Westcott, Peter W. 1964. Invasion of Clark Nutcrackers and Piñon Jays into southeastern Arizona. *Condor* **66**:441.

Whitmore, T. C. 1975. *Tropical Rain Forests of the Far East.* Clarendon Press, Oxford. 282 p.

Willson, Mary F., Helen J. Michaels, Robert I. Bertin, B. Benner, S. Rice, T. D. Lee, and A. P. Hartgerink. 1990. Intraspecific variation in seed packaging. *American Midland Naturalist* **123**:179–185.

Wilmore, Sylvia Bruce. 1977. *Crows, Jays, Ravens and Their Relatives.* David and Charles, Vancouver. 208 p.

Wold, Myron L., and Ronald M. Lanner. 1965. New stool shoots from a 20-year-old swamp-mahogany *Eucalyptus* stump. *Ecology* **46**:755–756.

Woodmansee, Robert G. 1977. Clusters of limber pine trees: a hypothesis of plant-animal coaction. *Southwestern Naturalist* **21**:511–517.

Yeaton, Richard I. 1983. The effect of predation on the elevational replacement of digger pine by ponderosa pine on the western slopes of the Sierra Nevada. *Bulletin of the Torrey Botanical Club* **110**:31–38.

Yli-Vaakuri, Paavo. 1954. Untersuchungen über organische Wurzelverbindungen zwischen Baümen in Kiefernbestanden. *Acta Forestalia Fennica* **60**:103–117.

Yoon, T.-H., K.-J. Im, E. T. Koh, and J.-S. Ju. 1989. Fatty acid compositions of *Pinus koraiensis* seed. *Nutrition Research* **9**:357–361.

Zavarin, Eugene. 1988. Taxonomy of pinyon pines. In M-F. Passini, D. Cibrian Tovar, and T. Eguiluz Piedra (compilers), *II Simposio Nac. sobre Pinos Piñoneros,* CEMCA, Mexico D.F., 29–39.

Zavarin, Eugene, Z. Rafii, L. G. Cool, and K. Snajberk. 1991. Geographic monoterpene variability of *Pinus albicaulis. Biochemical Systematics and Ecology* **19**: 147–156.

Zykov, I. V. 1953. The kedrovka in Siberian forests. Translated by Leon Kelso. *Priroda* **7**:112–114.

INDEX